Maths Skills

for A Level

Biology

James Penny

Nelson Thornes

Published in 2013 by:
Nelson Thornes Ltd
Delta Place
27 Bath Road
CHELTENHAM
GL53 7TH
United Kingdom

13 14 15 16 17 / 10 9 8 7 6 5 4 3 2 1

A catalogue record for this book is available from the British Library

ISBN 978 1 4085 2118 2

Page make-up and illustrations by Tech-Set Ltd, Gateshead
Printed and bound in Spain by GraphyCems

Contents

Acknowledgements

Photographs on pages 32–33 courtesy of iStockphoto
Table on page 79: Oikos: 'Annual bird ringing totals and population fluctuations', Christian Hjort and Claes-Göran Lindholm, © 1977. Reproduced with permission of Blackwell Publishing Ltd.

How to use this book

This workbook has been written to support the development of key mathematics skills required to achieve success in your A Level Science course. It has been devised and written by teachers and the practice questions included reflect the **exam-tested content** for AQA, OCR, Edexcel and Cambridge syllabi.

The workbook is structured into sections with each section having a clear scientific topic. Then, each spread covers a mathematical skill or skills that you may need to practise. Each spread offers the following features:

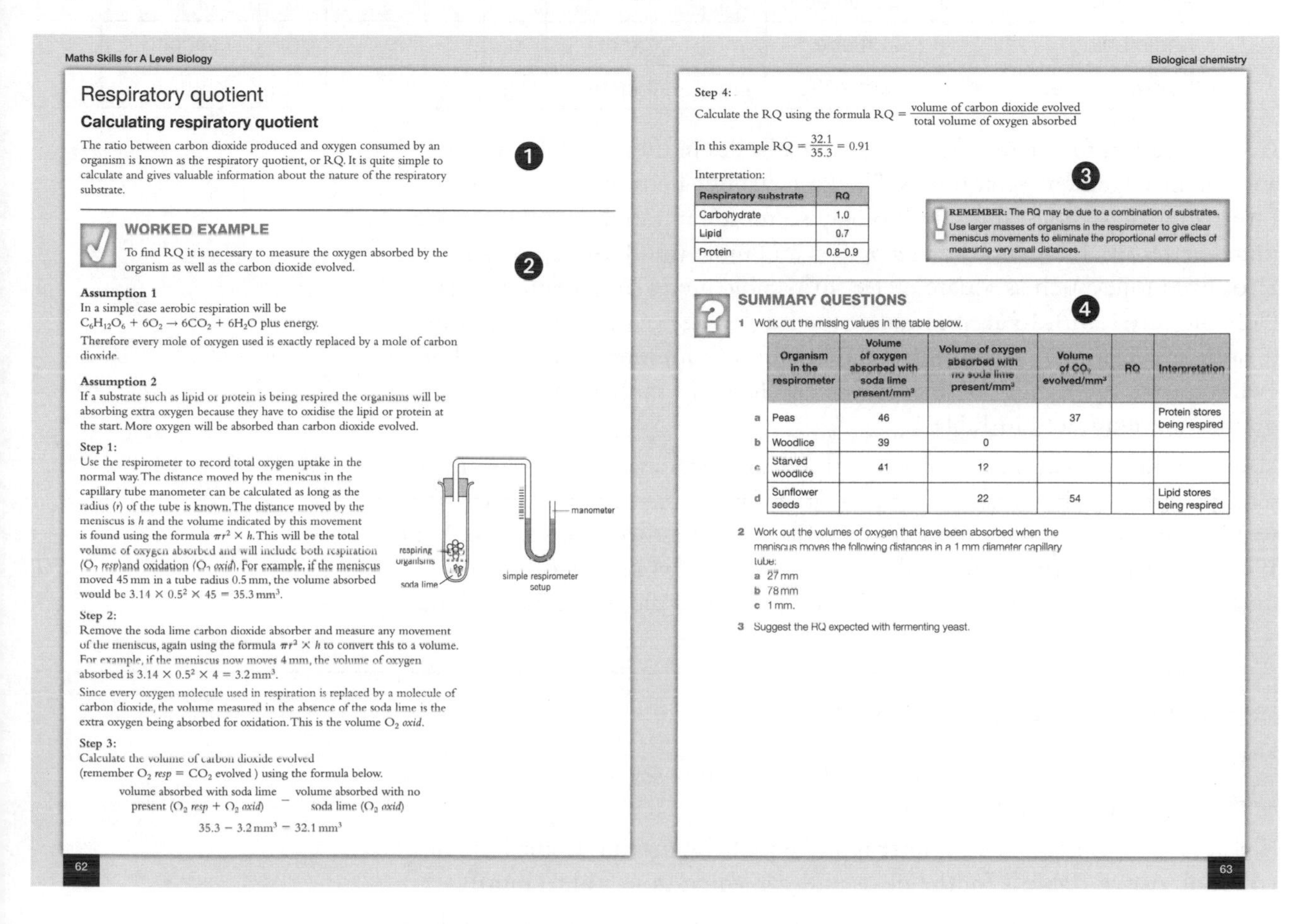
Maths Skills for A Level Biology

Biological chemistry

Respiratory quotient

Calculating respiratory quotient

The ratio between carbon dioxide produced and oxygen consumed by an organism is known as the respiratory quotient, or RQ. It is quite simple to calculate and gives valuable information about the nature of the respiratory substrate.

1

WORKED EXAMPLE

To find RQ it is necessary to measure the oxygen absorbed by the organism as well as the carbon dioxide evolved.

2

Assumption 1
In a simple case aerobic respiration will be
$C_6H_{12}O_6 + 6O_2 \rightarrow 6CO_2 + 6H_2O$ plus energy.

Therefore every mole of oxygen used is exactly replaced by a mole of carbon dioxide.

Assumption 2
If a substrate such as lipid or protein is being respired the organisms will be absorbing extra oxygen because they have to oxidise the lipid or protein at the start. More oxygen will be absorbed than carbon dioxide evolved.

Step 1:
Use the respirometer to record total oxygen uptake in the normal way. The distance moved by the meniscus in the capillary tube manometer can be calculated as long as the radius (r) of the tube is known. The distance moved by the meniscus is h and the volume indicated by this movement is found using the formula $\pi r^2 \times h$. This will be the total volume of oxygen absorbed and will include both respiration (O_2 *resp*) and oxidation (O_2 *oxid*). For example, if the meniscus moved 45 mm in a tube radius 0.5 mm, the volume absorbed would be $3.14 \times 0.5^2 \times 45 = 35.3\,mm^3$.

simple respirometer setup

Step 2:
Remove the soda lime carbon dioxide absorber and measure any movement of the meniscus, again using the formula $\pi r^2 \times h$ to convert this to a volume. For example, if the meniscus now moves 4 mm, the volume of oxygen absorbed is $3.14 \times 0.5^2 \times 4 = 3.2\,mm^3$.

Since every oxygen molecule used in respiration is replaced by a molecule of carbon dioxide, the volume measured in the absence of the soda lime is the extra oxygen being absorbed for oxidation. This is the volume O_2 *oxid*.

Step 3:
Calculate the volume of carbon dioxide evolved
(remember O_2 *resp* = CO_2 evolved) using the formula below.

volume absorbed with soda lime present (O_2 *resp* + O_2 *oxid*) − volume absorbed with no soda lime (O_2 *oxid*)

$35.3 - 3.2\,mm^3 = 32.1\,mm^3$

62

Step 4:
Calculate the RQ using the formula $RQ = \frac{\text{volume of carbon dioxide evolved}}{\text{total volume of oxygen absorbed}}$

In this example $RQ = \frac{32.1}{35.3} = 0.91$

Interpretation:

Respiratory substrate	RQ
Carbohydrate	1.0
Lipid	0.7
Protein	0.8–0.9

3

REMEMBER: The RQ may be due to a combination of substrates. Use larger masses of organisms in the respirometer to give clear meniscus movements to eliminate the proportional error effects of measuring very small distances.

SUMMARY QUESTIONS

4

1 Work out the missing values in the table below.

	Organism in the respirometer	Volume of oxygen absorbed with soda lime present/mm³	Volume of oxygen absorbed with no soda lime present/mm³	Volume of CO_2 evolved/mm³	RQ	Interpretation
a	Peas	46		37		Protein stores being respired
b	Woodlice	39	0			
c	Starved woodlice	41	12			
d	Sunflower seeds		22	54		Lipid stores being respired

2 Work out the volumes of oxygen that have been absorbed when the meniscus moves the following distances in a 1 mm diameter capillary tube:
a 27 mm
b 78 mm
c 1 mm.

3 Suggest the RQ expected with fermenting yeast.

63

❶ ***Opening paragraph*** outlines the mathematical skill or skills covered within the spread.

❷ ***Worked example*** – each spread will have one or two worked examples. The worked examples will be annotated.

❸ ***Remember*** is a useful box that will offer you tips, hints and other snippets of useful information.

❹ ***Summary questions*** are ramped in terms of difficulty and all answers are available at www.nelsonthornes.com

Numbers and units

Working with numbers and units

Table 1 Divisions of some units of measurement

Division	Prefix	Length		Mass		Volume		Time	
one thousand millionth	nano	nanometre	nm	nanogram	ng	nanolitre	nl	nanosecond	ns
one millionth	micro	micrometre	μm	microgram	μg	microlitre	μl	microsecond	μs
one thousandth	milli	millimetre	mm	milligram	mg	millilitre	ml	millisecond	ms
one hundredth	centi	centimetre	cm						
whole unit		metre	m	gram	g	litre	dm^3	second	s
one thousand times	kilo	kilometre	km	kilogram	kg				

A key criterion for success in biological maths lies in the use of correct units and the management of numbers. The units we use are from the Système Internationale – the SI units. In biology, we most commonly use the SI base units metre (m), kilogram (kg), second (s) and mole (mol). Biologists also use SI derived units, such as square metre (m^2), cubic metre (m^3), degree Celsius (°C) and litre (dm^3). To accommodate the huge range of dimensions in our measurements they may be further modified using appropriate prefixes. For example, one thousandth of a second is a millisecond (ms). Some of these prefixes are illustrated in Table 1.

When doing calculations, it's also important to express your answer using sensible numbers. For example, Mike worked out an answer of 6230 μm. It would have been more meaningful for Mike to express that answer as 6.2 mm. If you convert between units and round numbers properly it allows quoted measurements to be understood within the relevant scale of the observations.

WORKED EXAMPLE

To convert between units on the nano-, micro-, milli- and kilo- scale divide or multiply by 1000.

If you divide (to make the number more sensible by making it smaller), then you look **down** Table 1 for the next unit (e.g. going from μm to mm).

If you multiply (making a number bigger to make it more sensible) then look **up** Table 1 to the next unit (e.g. going from m to mm).

An exception is converting to centimetres. A centimetre is one hundredth rather than one thousandth of a metre.

For example:

a) to convert 0.006 dm^3 into millilitres, you multiply by 1000 to give 6 ml

b) to convert 6000 μg into milligrams, you divide by 1000 to give 6 mg

c) to convert 6000 m into km, you divide by 1000 to give 6 km.

Take care when using cubed units. A metre cubed means a cube with each side length 1 m or 1000 mm. The cube of 1000 is
$1000 \times 1000 \times 1000 = 1\,000\,000\,000$. So $1\ m^3 = 1\,000\,000\,000\ mm^3$.
Therefore, to convert between volumes expressed as cubed distances, your conversion factor is 1 000 000 000, rather than just 1000.

This means that:

a) $5\,000\,000\ mm^3$ is equivalent to $0.005\ m^3$

b) $6\,420\,000\ mm^3$ is equivalent to $0.00642\ m^3$

c) $0.000\,056\ m^3$ is equivalent to $56\,000\ mm^3$

Similarly, when converting between squared units, we need to do the same thing. For example, imagine converting from m^2 to mm^2. One square metre is $1000 \times 1000 = 1\,000\,000\ mm^2$. Therefore, to convert between areas, your conversion factor is 1 000 000, rather than just 1000.

WORKED EXAMPLE

Rounding

The rules for rounding are simple. Look at the figure to the right of the least significant figure you want to round to. If this figure is 5 or greater, round up. If this figure is less than 5, round down. For example:

a) 3.142 rounds to 3.14 (3 s.f.), rounds to 3.1 (2 s.f.) and rounds to 3 (1 s.f.).

b) 5.448 rounds to 5.45 (3 s.f.), rounds to 5.4 (2 s.f.) and rounds to 5 (1 s.f.).

> **REMEMBER:**
> **Significant figures**
> The first significant figure in a number is the first digit that is not zero. In 2.34 it is 2 and there are three significant figures; in 0.0056 it is 5 and there are two significant figures.

SUMMARY QUESTIONS

1 Undertake the following conversions:
- **a** 0.0062 mm into μm
- **b** 7928 ml into dm^3
- **c** 213 ml into dm^3
- **d** 4 000 000 ns into s
- **e** 727 m into km
- **f** 0.002 km into mm.

2 Undertake the following conversions:
- **a** $1\,000\,000\,000\ mm^3$ into m^3
- **b** $0.000\,001\ km^3$ into m^3
- **c** $0.000\,001\ m^3$ into mm^3.

3 Convert the following values so they make more sense to the reader. Choose the final units yourself. (Hint: make the final number as close in magnitude to zero as you can. For example, you would convert 1000 m into 1 km.)
- **a** 0.000 000 000 1 kg
- **b** 1 000 000 000 mg
- **c** $0.000\,000\,3\ dm^3$
- **d** 77 890 122 nm

4 Convert the following:
- **a** $1000\ mm^2$ into m^2
- **b** $0.6\ m^2$ into mm^2.

5 Round the following numbers:
- **a** 98.4478 to three significant figures
- **b** 1 298.444 444 4 to four significant figures
- **c** 5.555 55 to four significant figures
- **d** 0.358 to one significant figure
- **e** 0.000 464 8 to two significant figures.

> **REMEMBER:**
> **Write down the units!**
> When you do a calculation, it is very easy to forget to give the units. A number on its own makes no sense, unless the reader knows what the units are!
> Also, remember to put units only in headings in tables, **not** next to every figure entered.

> **REMEMBER:**
> **Units**
> It is common to use cm^3 in place of ml in biology. These units are in fact the same measurement. Occasionally cc is used to mean ml or cm^3.

The arithmetic mean

Calculating the arithmetic mean

For any range of observed measurements there should be a central or average value. This could simply be the middle value (the median) or it could be the most common value (the mode). The arithmetic mean, usually referred to simply as the mean, is a measure of central tendency that takes into account the number of times each measurement occurs as well as the range of the measurements. The greater the number of values averaged, the more precisely the mean will approach the true value, which will lie somewhere in the middle of the observed range. This is why it is sensible to repeat measurements, especially in biology where the natural unpredictability of living systems leads to inevitable variation in measurements.

WORKED EXAMPLE

The mean is determined by adding together all the observed values and then dividing by the number of measurements made.

For a range of values of x, the mean $\bar{x} = \frac{\sum x}{n}$

$\bar{x}$ is the mean value.

$\sum x$ is the sum of all values of x.

n is the number of values of x.

Example 1 Measuring fish

	Specimen 1	Specimen 2	Specimen 3	Specimen 4	Specimen 5
Fish length A/cm	45	57	39	72	51
Fish length B/cm	45.0	57.0	39.0	72.0	51.0

Example 2 Number of bubbles from pondweed per minute

Minute number	1	2	3	4	5
Number of bubbles per minute	21	30	27	41	28

In example 1, the mean fish length in sample A is:

$$(45 + 57 + 39 + 72 + 51) \div 5 = 52.8$$

So the average length of the fish in sample A is 52.8 cm. Notice that the mean gives us one extra decimal place. You need to be careful of this, because you cannot imply that measurements become more precise as a result of averaging. In this example the fish were measured to the nearest whole cm, so the mean should also be stated in this way. Rounding the calculated figure gives us a mean fish length of 53 cm for sample A. The fish in sample B are recorded to the nearest mm (note the decimal place). For this set the mean can be validly written as 52.8 cm.

In example 2, the mean number of bubbles per minute is:

$$(21 + 30 + 27 + 41 + 28) \div 5 = 29.4$$

In this case the introduction of the extra decimal place is not appropriate, as it is impossible to measure 0.4 of a bubble. The mean should be rounded and written as 29.

SUMMARY QUESTIONS

1 Find the mean values for the following sets of data. Justify the number of significant figures you quote each time.

   a Volumes of gas collected in 60 s from an enzyme reaction (cm^3): 62, 75, 65, 70, 68, 67

   b Number of birds visiting baits on a bird table per hour: 33, 31, 38, 32, 41, 24

   c Diameters of a sample of seeds (mm): 12.5, 14.0, 11.5, 11.0, 9.5

   d Number of bubbles produced by a sample of pondweed: 192, 186, 189, 178, 212

   e Lengths of mitochondria in a cell (μm): 1.8, 1.4, 0.9, 1.3, 1.6

2 Find mean values for the following odd sets of data. Select appropriate single units for each answer.

   a Times taken for an enzyme reaction to complete (minutes and seconds): 12:32, 11:56, 12:02, 14:00, 13:13

   b Distances moved by a sample of snails in five minutes: 1.6 m, 120 cm, 0.8 m, 960 mm, 2.3 m

   c Now find the mean speed of the snails in mm s^{-1}.

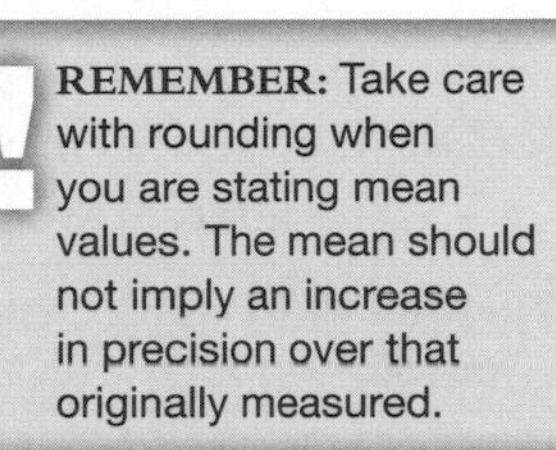

REMEMBER: Take care with rounding when you are stating mean values. The mean should not imply an increase in precision over that originally measured.

Arithmetic mean of grouped or tallied data

Calculating the arithmetic mean of grouped or tallied data

Biological data are often collected using some kind of tally system or by grouping into classes. This leads to tabulated data, which require a small adjustment to the method used to calculate the mean values. Tally data might include observations such as the number of bees visiting each colour of flower or the number of individuals born in each year. Grouped data could include the number of people in each age range or in each height range.

WORKED EXAMPLE

Example 1

Tally counts may be made of the number of turns made in a maze by different woodlice to give a table like this:

Number of turns possible = x	Number of woodlice with this number of turns = f	fx
1	3	3
2	5	10
3	11	33
4	16	64
5	8	40
6	7	42
7	1	7
	$\sum f = 51$	$\sum fx = 199$

Looking at columns 1 and 2 in the table, it is clear that there is a central tendency towards four turns. The formula for calculating the mean is:

$$\bar{x} = \frac{\sum fx}{\sum f}$$

Where:

$\bar{x}$ = the mean value

x = the values of each category of the independent variable, here the number of turns made

f = the frequency of the animals making each number of turns, i.e. how many were counted

$\sum$ = the sum of.

The calculation is made easier by the use of columns in the table.

Step 1: use column 3 to write in the values for fx (i.e. $f \times x$)

Step 2: add together the values in column 2 to find $\sum f$

Step 3: add together the values in column 3 to find $\sum fx$

Step 4: transfer the resulting numbers into the equation $\bar{x} = \frac{\sum fx}{\sum f}$

So in this example $\bar{x} = \frac{199}{51} = 4$ turns

Example 2

Data on body length of crabs is collected in classes.

Body length/mm	Mid points of each class = x	Number of crabs in this body length class = f	fx
20–25	22.5	1	22.5
25–30	27.5	5	137.5
30–35	32.5	9	292.5
35–40	37.5	2	75
40–45	42.5	2	85
		$\sum f = 19$	$\sum fx = 612.5$

In this example the crabs are in classes within a continuous variable, so the midpoint of the class is used as the value of x.

Step 1: use the same procedure as before to find the values of $\sum f$ and $\sum fx$.

Step 2: transfer the resulting numbers into the equation $\bar{x} = \frac{\sum fx}{\sum f}$

So in this example $\bar{x} = \frac{612.5}{19} = 32.2\,\text{mm}$

SUMMARY QUESTION

1 Calculate mean values for the following sets of data.

a

Number of bands on the shells of a sample of *Cepaea* snails	0	1	2	3	4	5
Number of snails of each band pattern	9	23	27	40	18	5

b

Leaf width of fern frond/mm	20–25	25–30	30–35	35–40	40–45
Number of leaves	12	25	13	5	2

c

Number of spots on the wing of a butterfly	0	1	2	3	4
Number of butterflies	3	7	16	46	11

d

Height jumped by fleas/mm	0–25	25–50	50–75	75–100	100–125
Number of fleas	23	45	19	8	4

Median and modal values

Using median and modal values

The arithmetic mean is the most familiar measure of central tendency, but sometimes it is more appropriate to use mode or median values to find the middle values in sets of data.

WORKED EXAMPLE

Median values

The median value in a set of data is calculated by placing the values in numerical order then finding the middle value in the range.

For example, the data set 12, 14, 11, 17, 9, 13, 13, 18, 10, 11 rearranges as 9, 10, 11, 11, 12, 13, 13, 14, 17, 18.

The middle of this range is 12.5

The median value is very useful when data sets have a few values at the extremes (outliers). If these values were included in a conventional mean they could skew the data. The median value also allows comparison of data sets with similar means but clear lack of overlap, skewed data and when there are too few measurements to calculate a reliable mean value.

For example, in the data set 1, 3, 10, 11, 12, 12, 12, 13, 14, 15 the median value is 12, a sensible looking midpoint, but the mean would be 10.3, skewed to the left by the numbers at the lower extreme.

Median values are useful in the Mann–Whitney *U* test, described later in this book.

WORKED EXAMPLE

Modal values

The modal value is very useful when examining data that is qualitative or in situations where the distribution has more than one peak (bimodal).

The modal value is the most frequent value in a set of data.

For example, in the data set 9, 10, 11, 11, 12, 13, 13, 13, 14, 17, 18, 19 the modal value is 13.

In biology, the sets of data may be small and this can introduce confusion. For example, in the data set 9, 10, 11, 11, 12, 13, 13, 14, 17, 18 there are apparently two modal values, 11 and 13, while in the set 11, 12, 13, 14, 17 there is no most frequent number and the mode is effectively every number and therefore of no value at all.

The modal value is not used very often.

Modal class

Where the data is categoric, for example, numbers of birds collecting baits of different colours, the modal class is the category (or categories) that has (have) the most data values.

REMEMBER: Use median values when your data has outliers to avoid skew and use a mode for categoric data to find the most common class.

SUMMARY QUESTIONS

1 Find the median values in the following data sets.

- a 23, 27, 28, 24, 32, 30, 31, 29, 26, 26
- b 127, 130, 199, 142, 175, 150, 131
- c 10, 6, 13, 16, 8, 7, 9, 12, 9, 11
- d 0.1, 3, 10, 4, 9, 10, 8, 3

2 Find the modal value of the following data sets.

- a 12, 14, 16, 20, 21, 21, 28, 30, 32, 34
- b 0.1, 0.9, 0.6, 0.8, 0.2, 0.6, 0.7
- c 4, 6, 5, 10, 7, 8, 4, 4

3 Find the modal class of the following data sets.

- a Number of insect pollinators on flowers: red 2; blue 45; white 22; yellow 41; purple 36
- b Shoe size: 38 – 3 people; 39 – 5 people; 40 – 6 people; 41 – 1 person; 42 – 2 people
- c Hair colour: black 2; brown 10; blonde 2; auburn 3; white 4
- d Flower count in quadrat: daisy 10; clover 15; buttercup 12; speedwell 15; dandelion 4

Percentages

Calculating percentages

A percentage is simply a fraction expressed as a decimal. It is an important thing to be able to calculate routinely, but is often incorrectly calculated in exams. These pages should allow you to practise!

WORKED EXAMPLE

Percentages as proportions

In a population the number of people who have brown hair was counted. The results showed that in the total population of 4600 people, 1800 people had brown hair.

The percentage of people with brown hair is found by calculating:

$$\frac{\text{number of people with brown hair}}{\text{total population}} \times 100$$

$$= \frac{1800}{4600} \times 100 = 39.1\%$$

Percentages as chance

In genetics predictions need to be expressed as a chance, which should always be a percentage.

Consider the monohybrid cross between two carriers of cystic fibrosis.

Parent	Male carrier		Female carrier	
Parent genotypes	Cc		Cc	
Possible gametes	C	c	C	c
F1 genotypes	CC	Cc	Cc	cc
F1 phenotypes	normal	normal (carrier)	normal (carrier)	sufferer

What is the chance of any baby born being a sufferer of cystic fibrosis?

Every birth carries a 1 in 4 chance so expressing this as a percentage:

$$\frac{\text{baby with cystic fibrosis}}{\text{all babies}} \times 100 = \frac{1}{4} \times 100 = 25\%$$

Percentage change

Percentage change is often used to describe osmosis experiments where samples (usually of potato tissue) gain and lose mass in different bathing solutions.

For example, a sample weighed 14.50 g at the start of the osmosis experiment and at the end it weighed 10.73 g.

The actual loss in mass = 14.50 g – 10.73 g = 3.77 g

$$\text{The percentage change} = \frac{\text{mass change}}{\text{starting mass}} \times 100 = -\frac{3.77}{14.50} \times 100 = -26\%$$

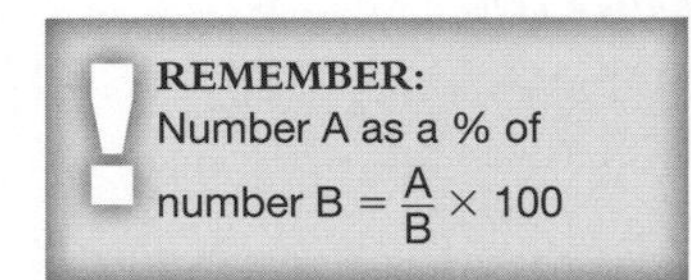

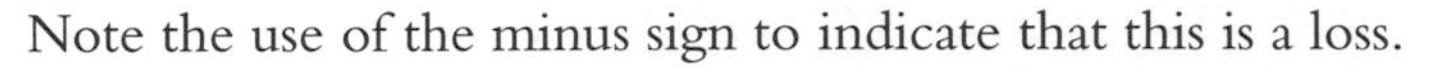

Note the use of the minus sign to indicate that this is a loss.

SUMMARY QUESTIONS

1 Calculate the values for the following situations.
   a The chance of an albino child being born to two heterozygous people (albinism is recessive).
   b The size of a population in which 67 of the people are left handed and with these people being 27% of the total.
   c The percentage of plantains in a sample of 670 plants of which 34 are plantains.

2 Convert the following fractions to percentages.
   a $\frac{1987}{10\,000}$
   b $\frac{45}{71}$
   c $\frac{3}{5}$
   d $\frac{4500}{10^6}$

3 Convert the following decimals to percentages (you should do this without a calculator).
   a 0.71
   b 0.34
   c 1.76

4 Convert the following common fraction ratios to percentages (you should do this without a calculator).
   a $\frac{1}{2}$
   b $\frac{1}{4}$
   c $\frac{2}{3}$

5 Convert the following percentages to fractions (you should do this without a calculator).
   a 75%
   b 33.3%
   c 25%

6 Convert the following mass changes to percentage changes and then plot a graph of % mass change against sucrose concentration.

Sucrose concentration/mol dm^{-3}	Initial mass/g	Final mass/g	Mass change/g	Percentage change in mass
0.1	1.82	2.55		
0.3	1.63	1.76		
0.5	1.95	1.70		
0.7	1.86	1.30		
0.9	1.79	1.06		

Decimals and standard form

Working with decimals and standard form

Sometimes biologists need to work with numbers that are very small, such as dimensions of organelles, or very large, such as populations of bacteria. In such cases the use of scientific notation or standard form is very useful, because it allows such numbers to be written easily.

WORKED EXAMPLE

Write down 63 900 000 000 as standard form.

Step 1 is to write down the smallest number between 1 and 10 that can be derived from the number to be converted. In this case it would be 6.39

Next write the number of times the decimal place will have to shift to expand this to the original number as powers of ten. On paper this can be done by hopping the decimal over each number like this:

6.3900000000

until the end of the number is reached.

In this example that requires 10 shifts, so the standard form should be written as 6.39×10^{10}.

For very small numbers the same rules apply, except that the decimal point has to hop backwards. For example 0.000 000 45 would be written as 4.5×10^{-7}.

So positive superscripts indicate the number of shifts forward and negative superscripts the number of shifts backwards.

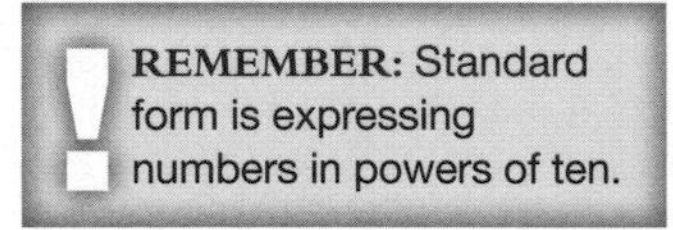

REMEMBER: Standard form is expressing numbers in powers of ten.

SUMMARY QUESTIONS

1 Convert the following numbers to standard form.

- **a** 100
- **b** 1000
- **c** 10 000
- **d** 0.1
- **e** 0.01
- **f** 0.001
- **g** 21 000 000
- **h** 435 000 000 000 000
- **i** 0.000 000 003 9

2 Write the following as decimals.

- **a** 10^6
- **b** 4.7×10^9
- **c** 1.2×10^{12}
- **d** 7.96×10^{-4}
- **e** 0.83×10^{-2}
- **f** 4.1×10^{-12}
- **g** 3.9×10^{-9}

3 Convert the following units to metres and write them in standard form.

a 1 mm

b 1 nm

c 1 μm

d 1 cm

e 27 mm

f 5647 mm

g 399 cm

h 29 000 000 μm

Data in line graphs

Presenting data in line graphs

The purpose of a line graph is to allow visualisation of a trend in a set of data. The graph can be used to make calculations, such as rates (see page 56) and also to judge the correlation between variables (see page 94). The certainty of the positions of the points can also be visualised by using confidence intervals (page 28). It is simple to draw such a graph but also quite easy to make simple mistakes.

WORKED EXAMPLE

Consider the set of colorimeter data below, collected from an experiment to investigate membrane damage and consequent pigment leakage from beetroot cells incubated at different temperatures.

Incubation temperature/°C	Mean % transmission of blue light by samples of beetroot pigment
20	74
30	68
40	69
50	36
60	16
70	12
80	11

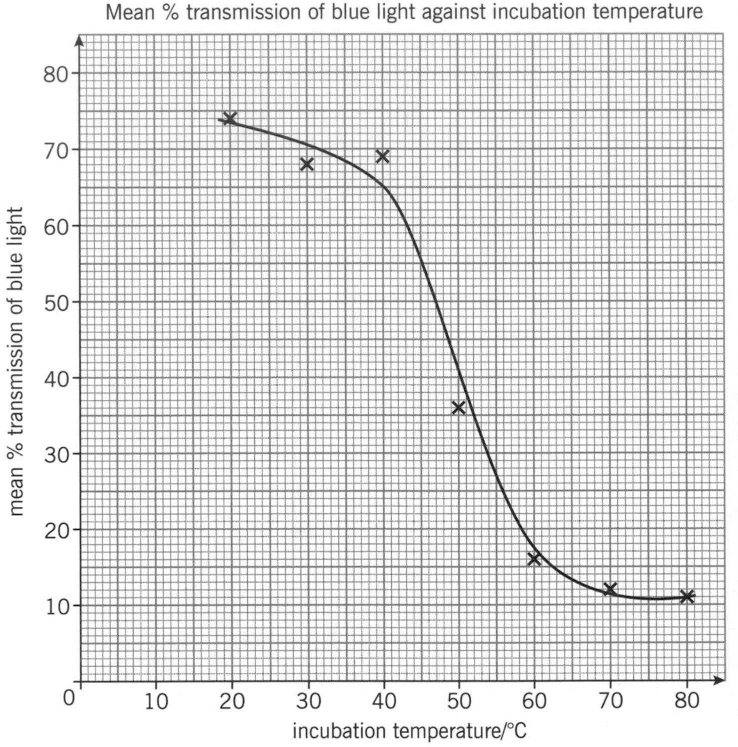

The rules when plotting the graph are:

- Ensure that the graph occupies the majority of the space available (in exams this means more than half the space).
- Mark axes using a ruler and divide them clearly and equidistantly (i.e. 10, 20, 30, 40 not 10, 15, 20, 30, 45.
- Ensure that the dependent variable that you measured is on the *y*-axis and the independent variable that you varied is on the *x*-axis.
- Ensure that both axes have full titles and units clearly labelled, e.g. pH of solution, not just 'pH'.
- Plot the points accurately using sharp pencil x marks so the exact position of the point is obvious.
- Draw a neat best fit line, either a smooth curve or a ruled line. It does not have to pass through all the points. Alternatively use a point to point ruled line, which is often used in biology where observed patterns do not necessarily follow mathematically predictable trends!
- Confine your line to the range of the points. Never extrapolate the line beyond the range within which you measured. Extrapolation is conjecture!
- Distinguish separate plotted trend lines using a key.
- Add a clear concise title.
- Where data ranges fall a long way from zero, a broken axis will save space. For example, if the first value on the *y*-axis is 36, it may be sensible to start the axis from 34 rather than zero. This will avoid leaving large areas of your graph blank.

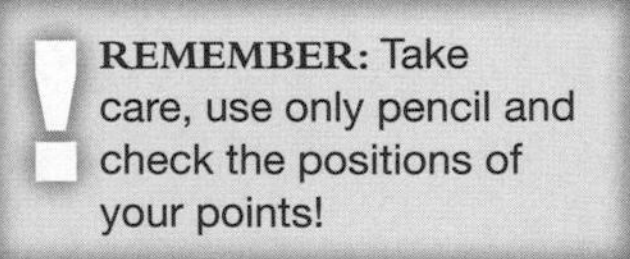

SUMMARY QUESTIONS

1 Plot suitable line graphs to illustrate the following sets of data. Use the graphs to answer the questions.

Turbidity of casein samples at different pH	
pH	**% transmission (blue light)**
9.00	99
8.00	99
6.00	87
5.00	67
4.75	26
4.50	30
4.00	24
3.75	43
3.50	64

Sucrose concentration /moles per litre	% change of mass of potato samples
0.9	−28.0
0.7	−16.7
0.5	−8.0
0.3	0.8
0.1	15.4
0.0	36.2

Sodium bicarbonate concentration /%	Rate of oxygen production by pondweed/$mm^3\ s^{-1}$
6.5	1.6
5.0	2.1
3.5	1.2
2.0	0.8
1.0	0.5
0.5	0.2

2 At the isoelectric point amino acids carry both positive and negative charge. Suggest the pH at which this occurs.

3 Estimate the sucrose concentration that is isotonic with the cell cytoplasm.

4 Suggest a possible optimum sodium bicarbonate concentration. How would you find this more precisely?

Exponential data in line graphs

Presenting exponential data in line graphs

Sometimes the rate of increase or decrease leads to rapidly increasing intervals apparent in results. Consequently, plotting a line graph can become difficult as the axes need to accommodate a range of values from very small to very large. In such circumstances a logarithmic graph may be the best solution.

WORKED EXAMPLE

Logarithmic data is often obtained from growth experiments with microorganisms such as bacteria. Asexual reproduction by binary fission, coupled with a short generation time, means that the population doubles every 20 minutes or so under ideal conditions. The example shows data collected from a culture of *E. coli*.

Time/min	Population as number of cells	Log_{10} number of cells
0	1	0.0
20	2	0.3
40	4	0.6
60	8	0.9
80	16	1.2
100	32	1.5
120	64	1.8
140	128	2.1
160	256	2.4
180	512	2.7

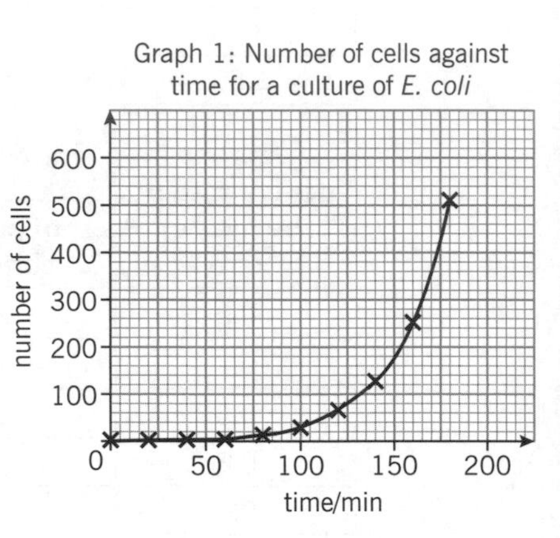

Graph 1: Number of cells against time for a culture of *E. coli*

Graph 1

Plotting the data on a conventional linear scale (graph 1) reveals the typical exponential curve. Two problems exist, firstly the difficulty in clearly dividing the *y*-axis to accommodate numbers such as 1 and 512 on the same scale and secondly the difficulty in using the graph to extrapolate in order to predict the population at a later time.

To make life easier, use a calculator to find the $\log_{10}$ values for each population count (these are in the table in column 3). Now plot the log data against time (see graph 2). If you are not sure how to do this, the calculator log function is explained on page 108.

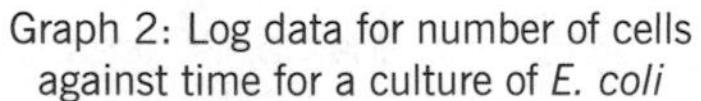

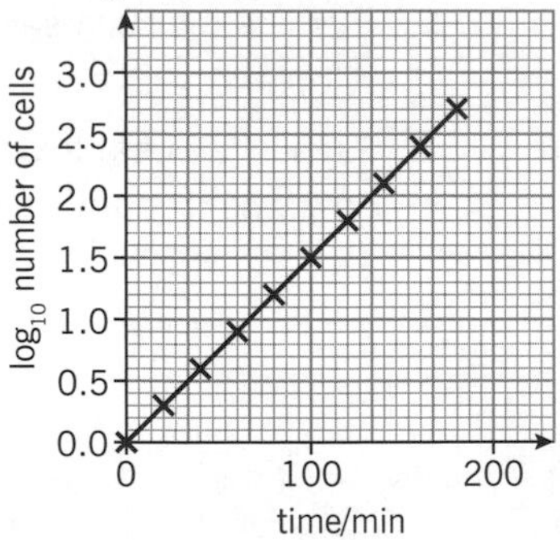

Graph 2: Log data for number of cells against time for a culture of *E. coli*

Graph 2

Notice that plotting the log data is much more useful, as the *y*-axis can now be scaled easily. Furthermore, the use of this method results in a straight line, which proves the logarithmic nature of the growth as well as providing scope for easy extrapolation and prediction of future populations.

REMEMBER: The use of logarithms ($\log_{10}$ or 'log to base ten') allows very large ranges of numbers to be easily accommodated on a simple axis by reducing the numbers to a scale between 0 and 10. For example, rather than try to scale an axis using divisions at 1, 10, 100 and 1000 you can use the logs of these numbers, which are 0, 1, 2 and 3 respectively. The logs express numbers in powers of 10.

To convert values you read off a $\log_{10}$ graph back to base 10

If you want to interpret data from a $\log_{10}$ graph, you may need to convert a value back to base 10. If you are unsure how to do this, the calculator function is explained at the top of page 109. For example, to determine the population at 150 minutes, read the y-value off the graph where time = 150. This is 2.25. Converting this value in $\log_{10}$ back to base 10 gives 178 (3 s.f.). If you check the table of data, you can see that this lies between the population values for 140 minutes and 160 minutes.

SUMMARY QUESTIONS

1 The table shows population counts of *Vibrio* bacteria after an initial inoculum at time = 0.

Time/min	Population/cells per cm^3
0	2 116 000
5	3 222 000
10	4 523 400
15	5 120 000
20	7 293 200
25	8 167 000
30	10 352 800
40	14 434 000
50	28 998 000
60	103 398 000

a Plot the data using a logarithmic scale for the population.
b Use your graph to estimate the likely population of cells after 27 minutes.
c Use extrapolation to work out the population expected after 120 minutes.

2 In an experiment growing yeast in a nutrient solution the cells were counted at hourly intervals. The following data were obtained.

Time/hours	Cell count/thousands per cm^3
0	500
1	600
2	720
3	870
4	1040
5	1250
6	1514

a Draw a graph to show the original population of cells against time.
b Attempt to use extrapolation to estimate the population of cells after 9 hours.
c Draw a logarithmic graph of the results.
d Use extrapolation to find the likely population of cells after 9 hours.
e Compare the two population estimates.

Normal distribution

Showing normal distribution

When quantitative data have been collected it is possible to draw a frequency distribution graph to illustrate the distribution of the data. Such graphs can be used to show the spread of the data around the mean value.

WORKED EXAMPLE

Sophie and Alison each measured the lengths of 125 woodlice. They recorded the number of animals in each of thirteen length classes. Note that they chose the class boundaries so there was no overlap. Each sampled the animals in a different population.

	Sophie's data	Alison's data
Length, ℓ mm	**Number of animals**	
$0 < \ell \leq 1.5$	2	3
$1.5 < \ell \leq 3.0$	3	7
$3.0 < \ell \leq 4.5$	5	9
$4.5 < \ell \leq 6.0$	10	11
$6.0 < \ell \leq 7.5$	15	12
$7.5 < \ell \leq 9.0$	19	13
$9.0 < \ell \leq 10.5$	24	14
$10.5 < \ell \leq 12.0$	17	13
$12.0 < \ell \leq 13.5$	12	12
$13.5 < \ell \leq 15.0$	9	11
$15.0 < \ell \leq 16.5$	6	9
$16.5 < \ell \leq 18.0$	2	7
$18.0 < \ell \leq 19.5$	1	4
n	125	125

To show the distribution of the animals in the different size classes at the two sites histograms should be plotted as shown on the next page.

In these examples, the mean lies in the middle category, which is often the case in a normal distribution, where the mean, median and mode are the same. This is only reliable if enough data has been collected and the data is in the form of numbers, rather than percentages.

In a set of normally distributed data 68% of the measurements of x lie within one standard deviation either side of the mean and 95% within two standard deviations. Thus the smaller the standard deviation, the more closely the data are clustered around the mean. Standard deviation is explained more fully on pages 24 and 25.

In both histograms 68% of the animals recorded lie within the range $x \pm 1s$, where s is the standard deviation. Sophie's animals show less spread, so the woodlice are clustered more closely around the mean, while Alison's woodlice show a wider spread and the calculated value of s was therefore larger.

> **REMEMBER:** Frequency data plotted in a histogram shows the spread of the data. The narrower the spread, the smaller the standard deviation of the data.

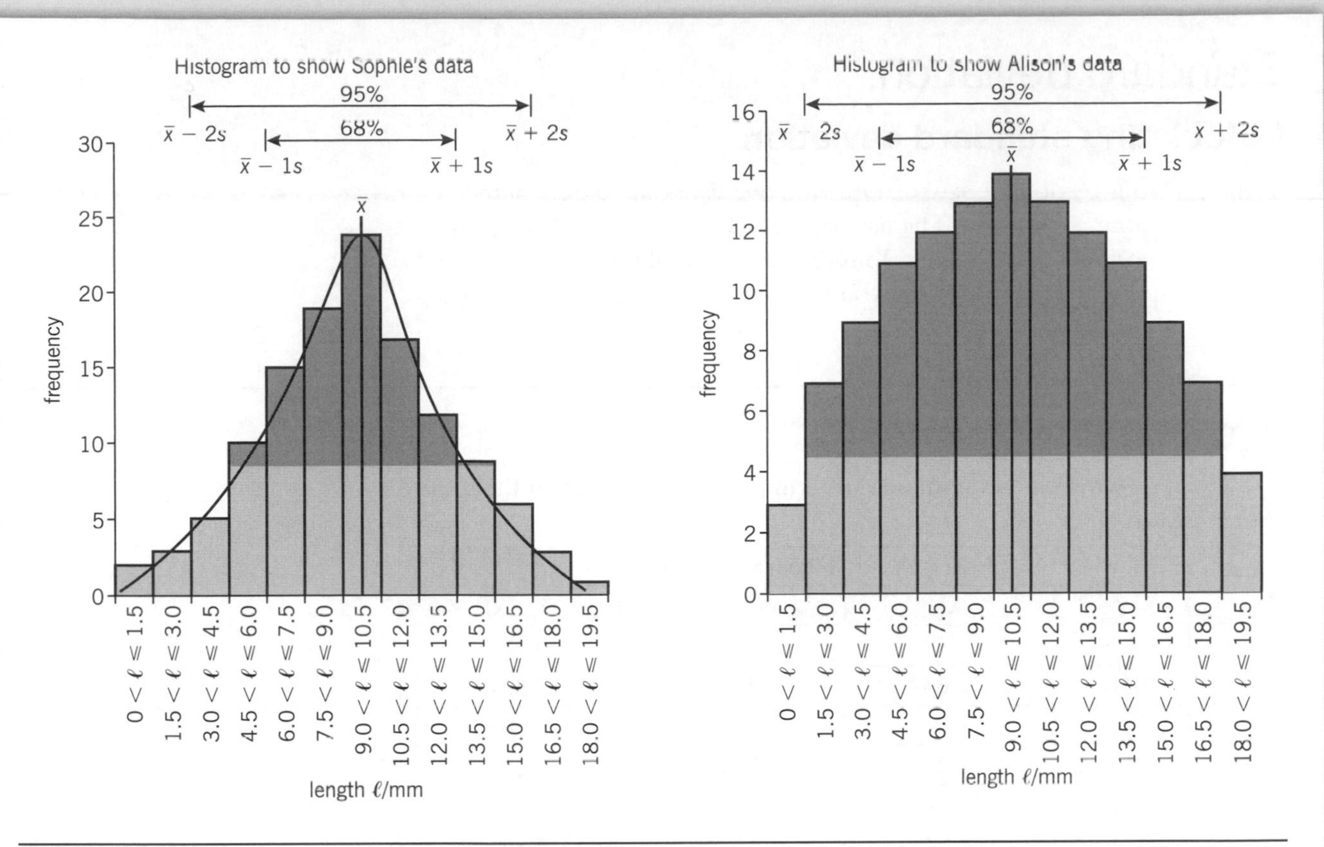

SUMMARY QUESTION

1 Plot a frequency distribution histogram for each of the following data sets. Which set of data has the widest spread?

Human height h/m	Number of people
$1.4 < h \leqslant 1.5$	30
$1.5 < h \leqslant 1.6$	50
$1.6 < h \leqslant 1.7$	120
$1.7 < h \leqslant 1.8$	160
$1.8 < h \leqslant 1.9$	110
$1.9 < h \leqslant 2.0$	40
$2.0 < h \leqslant 2.1$	20

Cockle shell diameter d/cm	Number of specimens
$1.2 < d \leqslant 1.4$	1
$1.4 < d \leqslant 1.6$	2
$1.6 < d \leqslant 1.8$	3
$1.8 < d \leqslant 2.0$	5
$2.0 < d \leqslant 2.2$	8
$2.2 < d \leqslant 2.4$	9
$2.4 < d \leqslant 2.6$	10
$2.6 < d \leqslant 2.8$	12
$2.8 < d \leqslant 3.0$	12
$3.0 < d \leqslant 3.2$	11
$3.2 < d \leqslant 3.4$	8
$3.4 < d \leqslant 3.6$	6
$3.6 < d \leqslant 3.8$	3
$3.8 < d \leqslant 4.0$	1
$4.0 < d \leqslant 4.2$	1

100 Dogwhelks: shell length / mm			
23	26	25	27
21	24	24	28
21	28	26	29
30	26	26	26
31	20	24	24
22	26	29	26
25	24	25	30
27	24	27	27
26	25	27	24
25	26	22	26
24	28	26	26
24	29	28	25
25	26	26	28
26	25	25	27
27	23	23	26
28	25	26	25
27	30	26	21
23	27	27	25
24	27	24	24
25	29	28	26
23	27	27	27
25	25	26	26
22	27	24	25
22	28	30	26
27	23	27	24

Standard deviation

Calculating standard deviation

In the previous section we saw that quantitative data can be distributed around a central mean value. The data may be clustered quite close to the mean, or it may be spread widely. To measure the variability in the data and judge the spread around the mean you will need to calculate the standard deviation.

WORKED EXAMPLE

Carey measured the masses of a sample of rose hips from the same bush.

Rose hip	Mass/g			
	x	x^2	$(x - \bar{x})$	$(x - \bar{x})^2$
1	1.6	2.56	−1.975	3.901
2	2.8	7.84	−0.775	0.601
3	4.2	17.64	0.625	0.391
4	3.5	12.25	−0.075	0.006
5	3.8	14.44	0.225	0.051
6	2.2	4.84	−1.375	1.891
7	5.0	25.0	1.425	2.031
8	4.9	24.01	1.325	1.756
9	4.0	16.0	0.425	0.181
10	2.9	8.41	−0.675	0.456
11	3.4	11.56	−0.175	0.031
12	3.0	9.0	−0.575	0.331
13	4.1	16.81	0.525	0.276
14	3.0	9.0	−0.575	0.331
15	3.4	11.56	−0.175	0.031
16	3.3	10.89	−0.275	0.076
17	5.4	29.16	1.825	3.331
18	3.6	12.96	0.025	0.001
19	4.3	18.49	0.725	0.526
20	3.1	9.61	−0.475	0.226
$n = 20$	$\sum x = 71.5$	$\sum x^2 = 272.03$		$\sum(x - \bar{x})^2 = 16.418$
	$\bar{x} = 3.575$			

There are different formulae for finding standard deviation, but they all do the same thing.

In **method 1**, the formula for standard deviation $s = \sqrt{\dfrac{\sum x^2 - \dfrac{(\sum x)^2}{n}}{n - 1}}$

x = the values measured, here rose hip mass

n = the number of values, here $n = 20$

$\sum$ = the sum of

Step 1: add together the values of x to give $\sum x$. This is shown in the table under column 2.

Step 2: find the squares of the values (x^2). These are shown in column 3.

Step 3: add together the values of x^2 to find $\sum x^2$. This is shown in the table under column 3.

Step 4: substitute the values into the equation to find s.

$$s = \sqrt{\frac{\sum x^2 - \frac{(\sum x)^2}{n}}{n-1}} \qquad s = \sqrt{\frac{272.03 - \frac{(71.5)^2}{20}}{20-1}} \qquad s = 0.93$$

> **NOTE:** Your calculator may offer two functions for working out the standard deviation: s (or σ_{n-1}) and σ. s is used to calculate the standard deviation of a sample whereas σ is used to calculate the standard deviation of a whole population. In biology you will normally be dealing with a sample so you should use the s (or σ_{n-1}) function.

In **method 2** the formula is $s = \sqrt{\frac{\sum(x-\bar{x})^2}{n-1}}$

Step 1: find the mean value of the values, $\bar{x}$. This is shown under column 2.

Step 2: find the values of $(x - \bar{x})^2$. These are shown in column 5 in the table.

Step 3: total up the values of $(x - \bar{x})^2$, as shown at the bottom of column 5.

Step 4: substitute the values into the equation to find s.

$$s = \sqrt{\frac{\sum(x-\bar{x})^2}{n-1}} \qquad s = \sqrt{\frac{16.418}{19}} \qquad s = 0.93$$

Statisticians tell us that in a set of normally distributed data 68% of the measurements of x will lie within one standard deviation either side of the mean and 95% within two standard deviations either side of the mean. The standard deviation can therefore be used on graphs and charts to show the limits between which 68% and 95% of the data fall, as shown on page 23.

SUMMARY QUESTIONS

1 Use method 1 to calculate means and standard deviation values for each concentration of carbon dioxide in the following set of data, which shows locust ventilation rates in different concentrations of carbon dioxide. Which row of data has the least spread and is therefore the most reliable?

Carbon dioxide/%	Ventilation rate/breaths per minute				
0	2	1	2	6	4
5	14	13	21	11	14
10	19	22	24	14	21
15	21	23	21	14	31
20	25	32	38	31	39

2 In an experiment the following data were collected from a respirometer using germinating peas at different temperatures. Use method 2 to calculate means and standard deviations for each temperature.

Temperature/°C	Meniscus movement in one minute/mm				
10	5	7	4	4	6
20	9	11	11	8	12
30	23	31	28	25	24
40	40	39	45	44	41
50	36	30	31	26	28

> **REMEMBER:** Standard deviation (SD) gives a measure of the spread of the data around a mean. Large values of SD show data to be less reliable.

Simple probability and normal distribution

Calculating simple probability and normal distribution

A population of values in a sample can be represented graphically as a normal distribution curve. Since the area under the curve represents the whole population, it is possible to work out the different proportions of the population within a specified range. This proportion can be represented as a probability.

WORKED EXAMPLE

In the normal distribution curve theoretical statistics tell us that 68% of the population will fall within the limits $\bar{x} \pm 1s$, where $\bar{x}$ = the mean value and s = the standard deviation, while 95% of the sample will lie within the range $\bar{x} \pm 2s$ (fig. 1).

These values can be expressed as proportions if the total population under the curve = 1 (100%). So 68% is represented as 0.68 and 95% as 0.95

These values are equivalent to probabilities; the probability of a value within the range $\bar{x} \pm 2s$ = 95%. The probability of a value being greater than $\bar{x} + 2s$ = 2.5%, as shown on the graph (fig. 2).

In terms of probability, it is therefore most likely that values will lie in the middle area of this plot and unlikely that values will lie further to the side of the mean value.

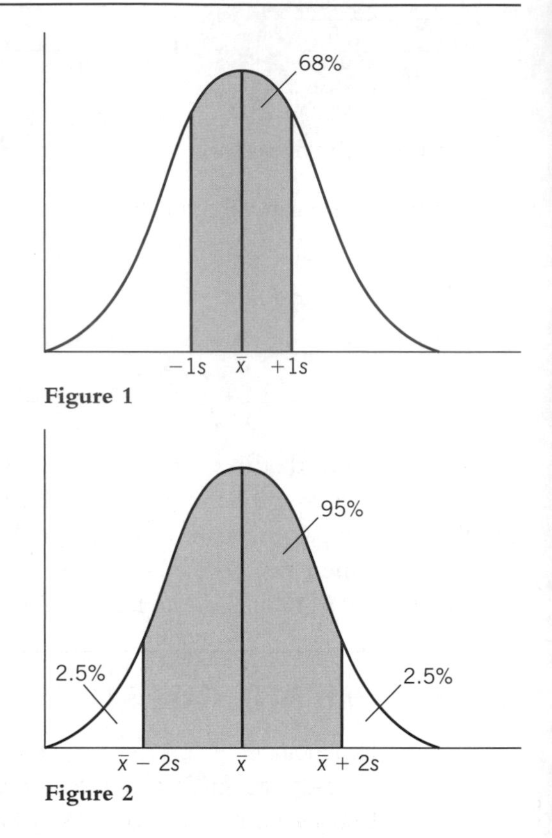

Figure 1

Figure 2

STRETCH YOURSELF!

It is possible to calculate actual values for probabilities.

To make calculations the values on the x-axis need to be represented using something called the standardised normal deviation, represented by the symbol z. The mean value is represented by $z = 0$ and lies at the midpoint of the curve.

To find proportions of the population under different parts of the curve the value of z is calculated and the table of values of z (see page 112) is applied to find the area under the curve.

Example 1: a population of rabbits was weighed and the mean mass was 2.5 kg with a standard deviation of 0.75. What is the probability that a rabbit will weigh more than 3 kg?

Step 1 is to find the value of z corresponding to $s = 0.75$ and mass 3 kg using the formula:

$z = \frac{x - \bar{x}}{s}$ so $z = \frac{3 - 2.5}{0.75} = 0.67$.

On the graph (fig. 3) the proportion of the population to the right of this is the proportion with a mass >3 kg. Using the table of standardised normal distribution (page 112) a value of $z = 0.67$ gives us a proportion of 0.2486 (to read the table look down column 1 to find 0.6 then across to the column headed .07 to find the value for 0.67).

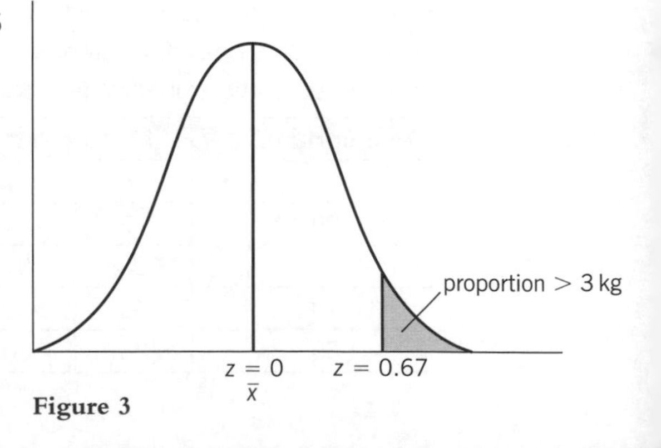

Figure 3

The area to the right of the mean is obviously only half the total population or 0.5. The area to the right of $z = 0.67$ is therefore $0.5 - 0.2486 = 0.2514$. This is approximately 0.25 or 25% of the rabbits in the population. The probability of a rabbit weighing more than 3 kg is therefore 25%.

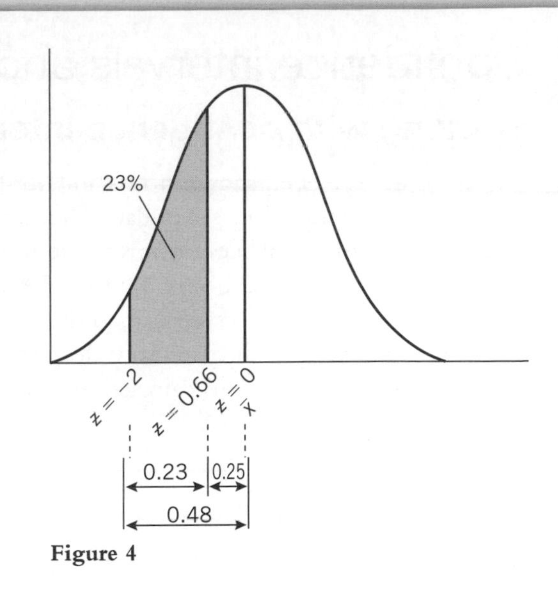

Figure 4

Example 2: What is the probability of a rabbit weighing between 1 and 2 kg?

Find the values of z for the two masses:

For 1 kg $z = \frac{x - \bar{x}}{s}$ so $z = \frac{1 - 2.5}{0.75} = -2.0$

For 2 kg $z = \frac{x - \bar{x}}{s}$ so $z = \frac{2 - 2.5}{0.75} = -0.67$

Now use the table of z values to find the proportions between 0 and 1 kg and between 0 and 2 kg. The proportion between 1 and 2 kg will be the difference between these.

$z = -2$ gives a proportion of 0.4772.
$z = -0.67$ gives a proportion of 0.2486.

The difference between the two is $0.4772 - 0.2454 = 0.2318$, which rounds to 0.23.

So the probability of a rabbit weighing between 1 and 2 kg in this sample population is 23% (fig. 4).

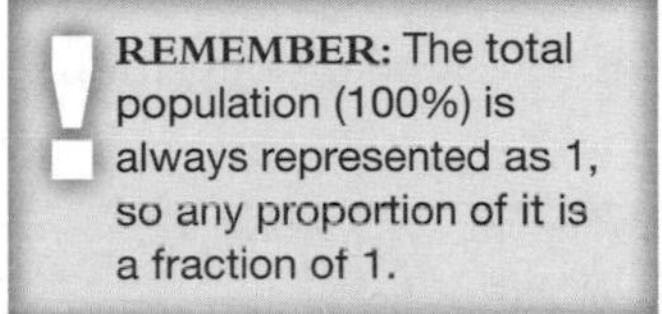
REMEMBER: The total population (100%) is always represented as 1, so any proportion of it is a fraction of 1.

SUMMARY QUESTIONS

1 The length of horns was measured in a large sample of highland cattle. The mean length of horn was 59 cm and the standard deviation was 11.9. The sample may be assumed to be normally distributed. What is the probability of cattle having horns:

 a >70 cm long
 b <35 cm long
 c between 80 and 90 cm long?

2 In a sample of leaves the probability of any leaf being more than 160 mm wide is 4.6%. If the standard deviation of the original sample was 11.8 what is:

 a the mean leaf diameter of the sample if the value of z at 95.4% is 1.685. (Hint: you will need to rearrange the equation for z from the opposite page to find $\bar{x}$.)
 b the probability of a leaf being between 120 and 130 mm wide? (Hint: Start by drawing yourself a sketch of the normal curve and mark on the value of the mean ($\bar{x}$) from part **a** and the values from this question part. Then, using $\bar{x}$ and the value of s from the question stem, work out the values of z for 120 mm and 130 mm. Use the table on page 112 to find the corresponding values – you can ignore the minus signs since both values are on the same side of the mean – and subtract one from the other to get the probability of a leaf having a width between 120 and 130 mm.)

Confidence intervals and reliability of the mean

Working with confidence intervals and reliability of the mean

You have already been shown how standard deviation can be used to indicate the amount of spread in a set of data. The more spread the data then the less certain it becomes that the mean is a true measure of the central tendency. For example, on page 22 both Sophie and Alison found the same mean, but Sophie's data had much less spread, so she had greater confidence in her mean. It is possible to calculate how close the mean of the sample is to the true mean and then use this calculation to plot confidence intervals on graphs and charts.

WORKED EXAMPLE

The first calculation is called the standard error (SE) of the mean. It is a numerical estimate of how close your sample mean is to the true mean.

The formula for standard error is $SE = \frac{s}{\sqrt{n}}$

s is the standard deviation.

n is the sample size.

Temperature/°C	Time taken to collect 10 cm^3 of gas/s								
	1	2	3	4	5	mean	*s*	SE	95% CI
15	87	95	102	121	117	104	14.4	6.44	±17.9
20	67	78	61	90	86	76	12.3		
25	57	59	48	66	51	56	7.0		
30	47	45	39	42	21	39	10.4		
35	118	123	145	136	132	131	10.7		

Table 1

Table 1 shows the data collected from an experiment collecting the gas from photosynthetic water plants at different temperatures. To find the standard error at 15 °C substitute the values into the formula as follows:

$$SE = \frac{s}{\sqrt{n}} \quad \text{so} \quad SE = \frac{14.4}{\sqrt{5}} = 6.4399$$

SE is often plotted directly onto graphs as error bars, but it is better to calculate proper 95% confidence intervals, which is quite easily done.

The formula required is: 95% confidence interval = SE × t

't' is a statistical value derived from the normal distribution. It depends on the sample size and linked with critical probability values of 5%. The t value is found from the standard table (Table 2 on page 29).

You will notice that in the table of t values as the sample size **increases** t gets **smaller**, so a larger sample will give a smaller confidence interval (CI).

To use the table find the appropriate degrees of freedom using the formula df = $(n - 1)$.

As there are 5 repeats the degrees of freedom is $5 - 1 = 4$

Table of values of t when $p = 0.05$							
Degrees of freedom (df = $n - 1$)	t	df	t	df	t	df	t
1	12.71	11	2.2	21	2.08		
2	4.3	12	2.18	22	2.07		
3	3.18	13	2.16	23	2.07		
4	2.78	14	2.15	24	2.06	40	2.02
5	2.57	15	2.13	25	2.06	60	2
6	2.45	16	2.12	26	2.06	120	1.98
7	2.37	17	2.11	27	2.05	>120	1.96
8	2.31	18	2.1	28	2.05		
9	2.26	19	2.09	29	2.05		
10	2.23	20	2.09	30	2.04		

Table 2

From table 2, at 4 degrees of freedom t =2.78

To find the 95% confidence interval of the mean at 15 °C substitute the numbers into the formula

$$95\%\ \text{CI} = \text{SE} \times t \text{ so } 95\%\ \text{CI} = 6.4399 \times 2.78 = 17.9$$

This value can now be plotted as error bars on the graph as a bar extending ±17.9 either side of the plotted point for the mean, as shown below. This is a visual indication of the range within which 95% of the points lie within 2 standard deviations either side of the mean, which are the limits normally acceptable in biology (see page 24). The shorter the error bar the smaller the spread in the data and the more certain the position of the plotted point. For bar charts plot the error bar as a vertical line through the centre of the top of the bar.

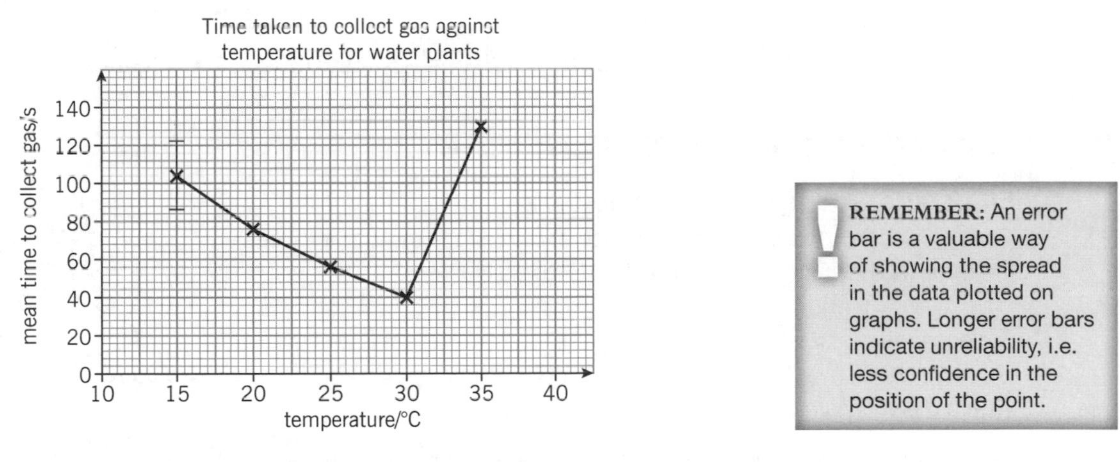

REMEMBER: An error bar is a valuable way of showing the spread in the data plotted on graphs. Longer error bars indicate unreliability, i.e. less confidence in the position of the point.

SUMMARY QUESTIONS

1. Calculate the standard error for the remaining four rows in table 1.
2. Use the values of SE to find the 95% confidence intervals for each of the remaining four means. Plot the error bars on a neat copy of the graph. Which is the most certain point? Which point could do with some further repeats to boost confidence in its position?

Surface areas and volumes

Calculating surface areas and volumes

The relationship between surface area and volume is of fundamental importance in biology. The ratio between the two governs the exchange of substances and energy and consequently governs the efficiency of biological design. Living organisms can often be approximated to geometrical shapes, allowing estimation of surface areas and volumes.

WORKED EXAMPLES

Surface areas of three-dimensional objects

Cubes and cuboids (fig. 1)

Surface area is $2(ab) + 2(ac) + 2(bc)$

If $a = 4$ cm, $b = 2$ cm and $c = 6$ cm the surface area is

$$2(4 \times 2) + 2(4 \times 6) + 2(2 \times 6) = 88 \text{ cm}^2$$

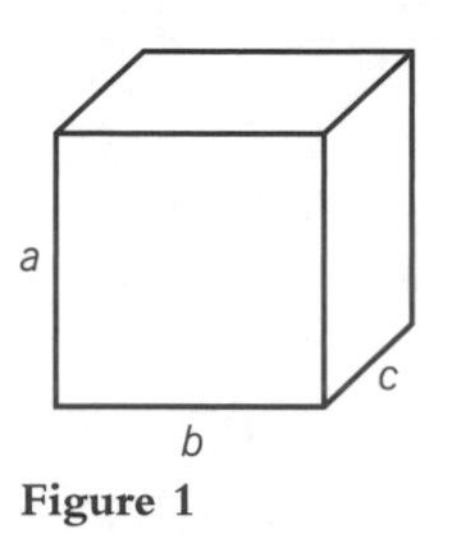

Figure 1

Spheres (fig. 4)

Surface area is $4\pi r^2$

If $r = 4$ cm the surface area is $4 \times 3.14 \times 4^2 = 201 \text{ cm}^2$

Cylinders (fig. 2)

Surface area is $2\pi r^2 + (\pi d \times h)$

Where d = diameter and h = height (or length)

If $d = 4$ cm and $h = 10$ cm the surface area is

$$2 \times 3.14 \times 2^2 + (3.14 \times 4 \times 10) = 25.1 + 125.6 = 151 \text{ cm}^2 \text{ (3 s.f.)}$$

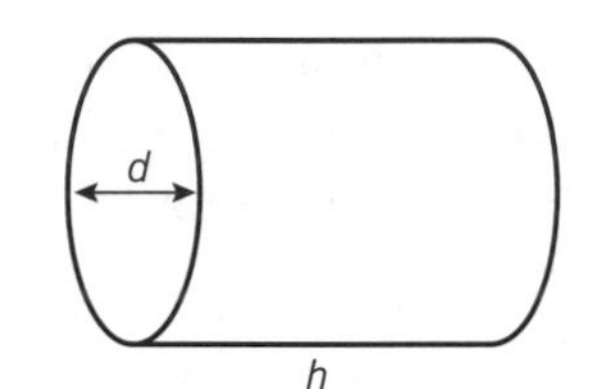

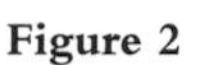

Figure 2

STRETCH YOURSELF

Ellipsoids – complex but useful!

This calculation is really an approximation and uses a formula that relates the three radii a, b and c in a general 3D ellipsoid (fig. 3).

$$\text{Surface area} = 4\pi\left(\frac{(a^pb^p + a^pc^p + b^pc^p)}{3}\right)^{\frac{1}{p}}$$

Where p is a constant ≈ 1.6075

If $a = 2$ cm, $b = 4$ cm and $c = 6$ cm the surface area is

$$= 4 \times 3.14\left(\frac{(2^{1.6075}4^{1.6075} + 2^{1.6075}6^{1.6075} + 4^{1.6075}6^{1.6075})}{3}\right)^{\frac{1}{1.6075}}$$

$$= 12.56 \times \left(\frac{(3.047 \times 9.286) + (3.047 \times 17.819) + (9.286 \times 17.819)}{3}\right)^{0.6221}$$

$$= 12.56 \times \left(\frac{28.295 + 54.295 + 165.467}{3}\right)^{0.6221}$$

$$= 12.56 \times 82.685^{0.6221}$$

$$= 12.56 \times 15.588$$

$$= 196 \text{ cm}^2 \text{ (3 s.f.)}$$

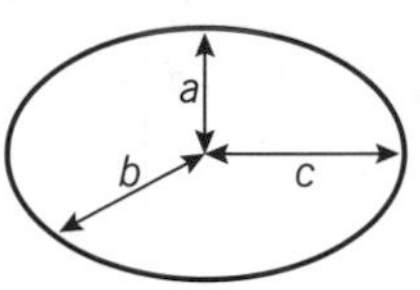

Figure 3

REMEMBER: To do the power functions on a calculator, press the number to multiply, then press the power button y^x then press the desired power. E.g. to find $3^{1.4}$ enter 3, then press y^x then enter 1.4; you should get the answer 4.66

WORKED EXAMPLES

Volumes of three dimensional objects

Cubes and cuboids (fig. 1)

Volume $= a \times b \times c$

If $a = 4$ cm, $b = 2$ cm and $c = 6$ cm the volume $= 4 \times 2 \times 6 = 48\ \text{cm}^3$

Spheres (fig. 4)

Volume $= \frac{4}{3}\pi r^3$

For a sphere radius 4 cm the volume $= \frac{4}{3} \times 3.14 \times 64 = 268\ \text{cm}^3$ (3 s.f.)

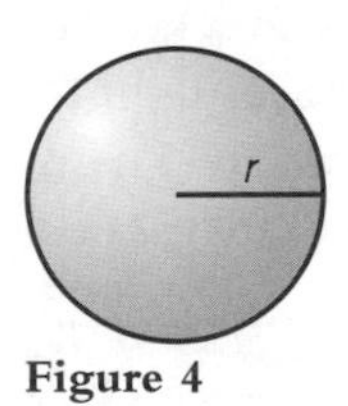

Figure 4

STRETCH YOURSELF

Ellipsoids again!

Volume $= \frac{4}{3}\pi\,(r_1 \times r_2 \times r_3)$ (modifies the volume of a sphere to take into account the different radii). r_1, r_2 and r_3 being the three radii labelled a, b and c on fig. 3.

For the ellipsoid $a = 2$ cm, $b = 4$ cm and $c = 6$ cm the volume

$= \frac{4}{3} \times 3.14\ (2 \times 4 \times 6)$
$= 4.19 \times 48$
$= 201\ \text{cm}^3$ (3 s.f.)

Cylinders (fig. 2)

Volume $= \pi r^2 h$

For the cylinder $d = 4$ cm ($r = 2$ cm) and $h = 10$ cm so the volume $= 3.14 \times 2^2 \times 10 = 126\ \text{cm}^3$ (3 s.f.)

WORKED EXAMPLE

Surface area : volume ratios

Surface area : volume ratio is found by dividing the surface area (SA) by the volume of the object.

For example, the cube considered above has a SA : volume ratio of $88 \div 48 = 1.8$

The sphere considered above has a SA : volume ratio of $201 \div 268 = 0.75$

The ellipsoid has a SA : volume ratio of $186.1 \div 200.96 = 0.93$

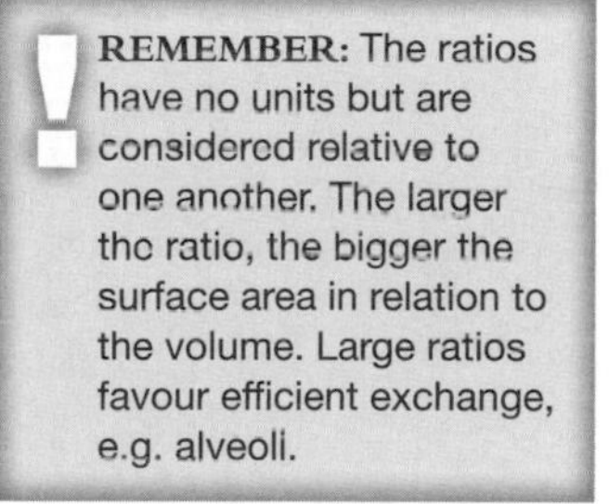
REMEMBER: The ratios have no units but are considered relative to one another. The larger the ratio, the bigger the surface area in relation to the volume. Large ratios favour efficient exchange, e.g. alveoli.

SUMMARY QUESTIONS

1 Estimate the surface area to volume ratios of cubes of side lengths
  a 2 cm **b** 4 cm **c** 6 cm

2 **a** Estimate the surface area and volume of a lung, assuming it is a cylinder of length 30 cm and diameter 15 cm
  b What is the surface area to volume ratio of this lung?
  c Estimate the surface area of a spherical alveolus diameter 200 μm
  d If the lung 30 × 15 cm has 480 million alveoli, estimate:
    i the total exchange surface area
    ii the surface area to volume ratio.

3 Estimate the surface area to volume ratios of eggs of the following dimensions then suggest which egg can be left longer by a feeding adult without it cooling down too much.
  a Length 9 cm, width 5 cm **b** Length 1 cm, width 0.8 cm

Magnification

Introduction to magnification

To look at small biological specimens you use a microscope to magnify the image that is observed. The microscope was developed in the 17th century. Anton van Leeuwenhoek used a single lens and Robert Hooke used two lenses. The lenses focus light from the specimen onto your retina to produce a magnified virtual image. The magnification at which observations have been made depends on the lenses used.

WORKED EXAMPLE

Lenses each have a magnifying power, defined as the number of times the image is larger than the real object. The magnifying power is written on the lens.

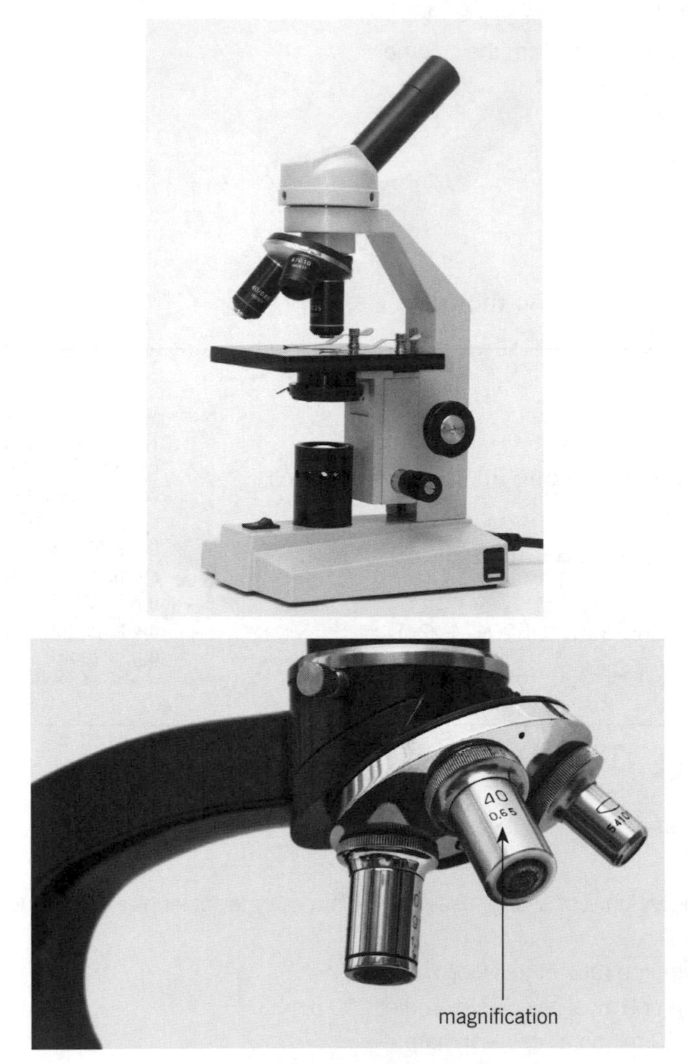

The magnification power of an objective or eyepiece lens is written on it

To find the magnification of the virtual image that you are observing, multiply the magnification powers of each lens used. For example, if the eyepiece lens is ×10 and the objective lens is ×40 the total magnification of the virtual image is 10 × 40 = 400.

REMEMBER: This is the virtual magnification as seen by the eye. It is not the same as the magnification of a printed image of the specimen.

SUMMARY QUESTION

1 Calculate the magnification of the virtual image produced by the following combinations of lenses:

- **a** objective ×10 and eyepiece ×12
- **b** objective ×40 and eyepiece ×15
- **c** objective ×100 and eyepiece ×12
- **d** objective ×20 and eyepiece ×10
- **e** objective ×100 and eyepiece ×20.

Graticules and stage micrometers

How to use a graticule and stage micrometer

To measure something under the microscope, you need to use a graticule, which is a fixed scale inside the eyepiece (fig. 1a). The divisions on the graticule can be used to measure magnified images, but you need to calibrate the graticule for each objective lens using a stage micrometer. This is a 10 mm scale etched onto a slide with subdivisions of precisely 0.1 mm.

WORKED EXAMPLE

Calibrating an eyepiece graticule

Place the stage micrometer (on the slide) on the microscope stage under low power and align it so you can see both the eyepiece graticule scale and the focused stage micrometer scale (fig. 1b). The stage micrometer will be the larger of the two scales.

- In fig. 1b, 10 eyepiece units are equivalent to one stage micrometer division of 0.1 mm.
- This means that 1 eye piece unit is equivalent to $\frac{0.1}{10} = 0.01$ mm
- This is equivalent to $0.01 \times 1000 = 10\ \mu m$

This process needs to be repeated when the objective lens is changed.

- At high magnification (fig. 1c) 26 eyepiece units are equivalent to one stage micrometer division of 0.1 mm.
- This means that 1 eyepiece unit is equivalent to $\frac{0.1}{26} \times 1000 = 3.9\ \mu m$

In fig. 1d, the micrometer has been replaced by the specimen and the objective lens is the one used in fig. 1b.

- The specimen is approximately 58 eyepiece units wide.
- Each eyepiece unit is 10 μm.
- So the length of the specimen is $58 \times 10\ \mu m = 580\ \mu m$

REMEMBER:
Use the correct calibration
Calibrate the length of one eyepiece unit under each different objective lens. Use the correct value when you start measuring!

SUMMARY QUESTIONS

1. A student calibrated a microscope with a ×10 objective lens and a ×10 eyepiece. The student found that 20 eyepiece graticule units were equivalent to 1 stage micrometer unit of 0.1 mm. Calculate the length of 1 eyepiece graticule unit.
2. A student calibrated a microscope and found that 10 eyepiece graticule units were equivalent to 2 stage micrometer units (each of which was 0.1 mm long). The student observed a specimen to be 7 eyepiece units long. What was the length of the specimen in micrometres?
3. When the student recalibrated the microscope using a high power objective lens, she found that 40 eyepiece graticule units were equivalent to 2 stage micrometer units (each of which was 0.1 mm long). The student viewed a specimen under these conditions, and observed it to be 4 eyepiece units long. What was the length of the specimen in micrometres?

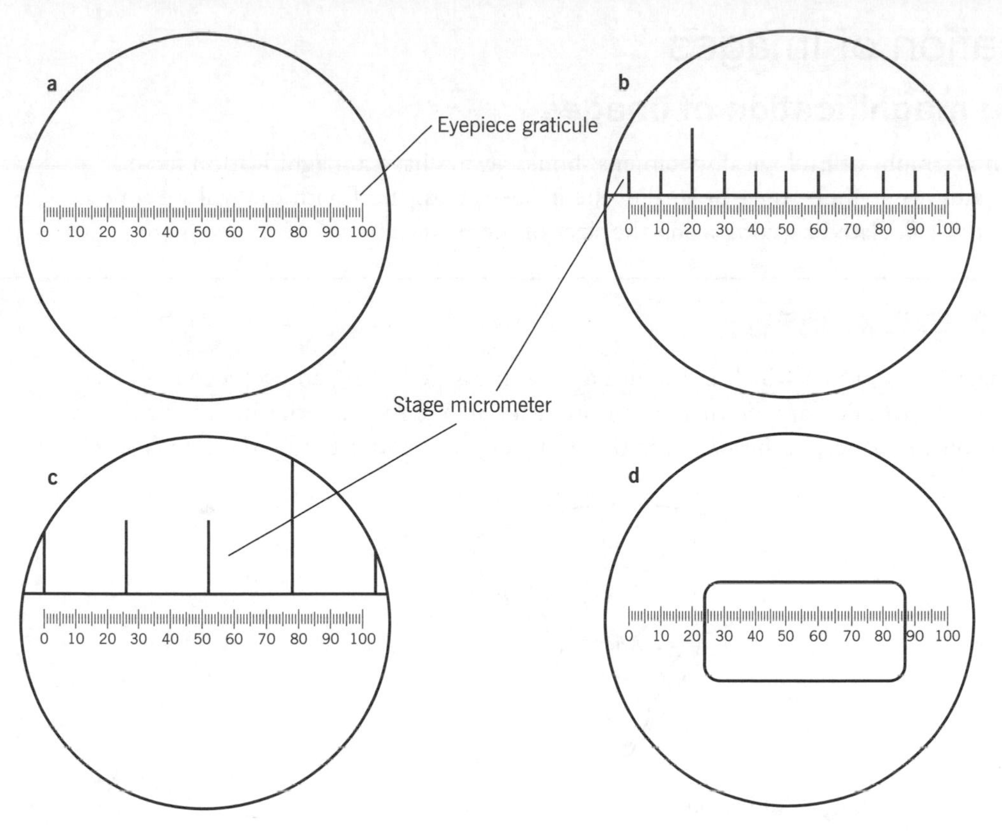

Figure 1

4 A student calibrated a microscope and found that 21 eyepiece graticule units were equivalent to 1 stage micrometer unit of 0.1 mm long. The student viewed a specimen under these conditions, and observed it to be 65 eyepiece units long. What was the length of the specimen in micrometres?

5 Calculate the values of the following:
- **a** 0.25 μm expressed in mm
- **b** 0.25 mm expressed in μm
- **c** 21 eyepiece graticule units expressed in μm, when 4 eyepiece graticule units are equivalent to 1 μm
- **d** 30 eyepiece graticule units expressed in μm, when 10 eyepiece graticule units are equivalent to 5 μm
- **e** 7 eyepiece graticule units expressed in mm, when 3 eyepiece graticule units are equivalent to 0.5 μm.

Magnification of images

Calculating magnification of images

Drawings and photographs of biological specimens should always have a magnification factor stated. This indicates how much larger or smaller the image is compared with the real specimen. The magnification is calculated by comparing the sizes of the image and the real specimen.

WORKED EXAMPLE

The image shows a flea which is 1.3 mm long as measured using an eyepiece graticule (see page 34). To calculate the magnification of the image measure the image (or the scale bar if given) on the paper, in this example the body length as indicated by the line A – B.

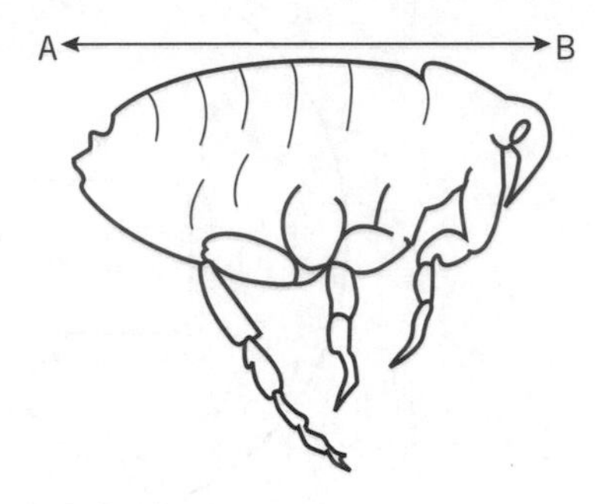

$$\text{Magnification} = \frac{\text{length of image}}{\text{length of real specimen}}$$

For this image the length of the image is 42 mm and the length of the real specimen is 1.3 mm.

The magnification is therefore $42 \div 1.3 = 32.31$

The magnification factor should therefore be written as $\times 32.31$

SUMMARY QUESTION

1 For each of the images shown, calculate a suitable magnification factor.

a A mitochondrion that is 1.5 μm long.

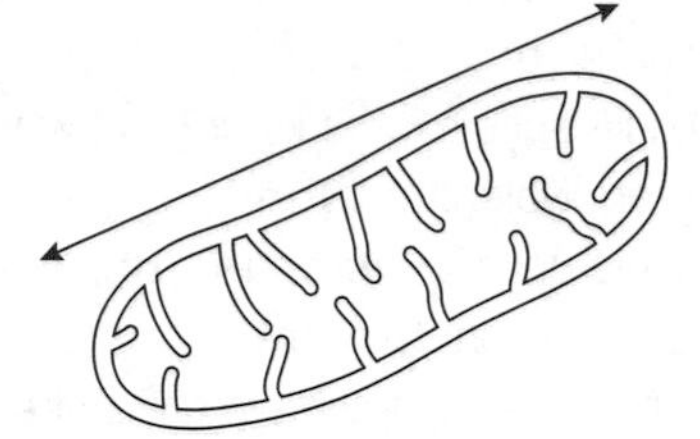

b A pollen grain that is 50 μm wide.

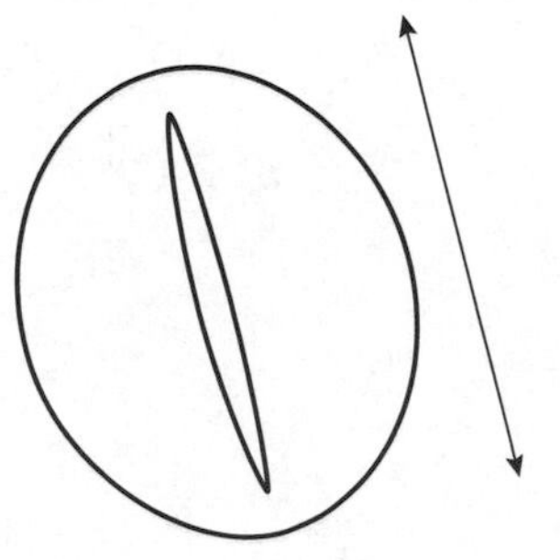

c Villi in the small intestine that are 1.2 mm long from A to B.

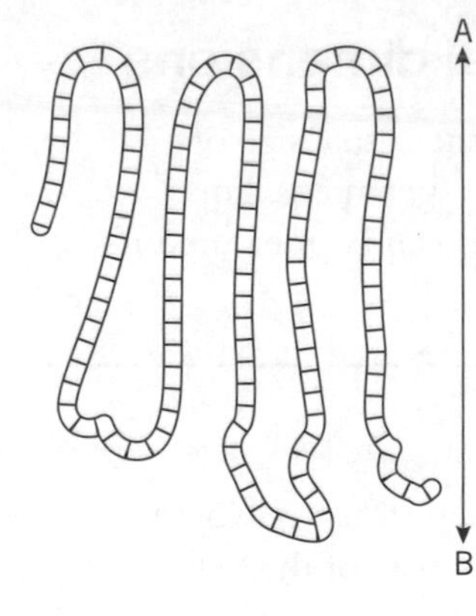

d An African elephant (the scale bar is 1.0 m).

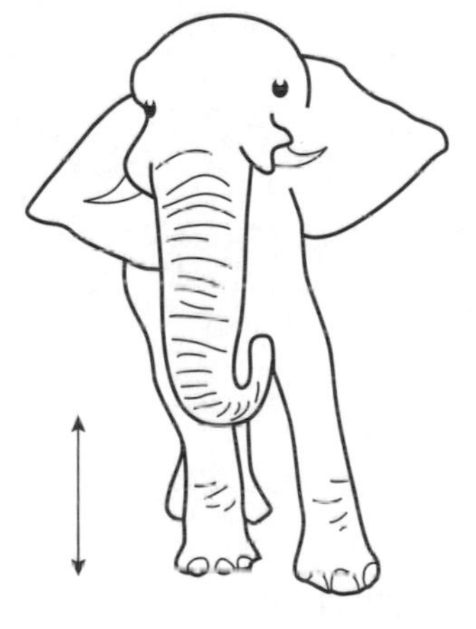

REMEMBER:

Use the same units

A common error is to mix units when performing these calculations. Begin each time by converting measurements to the same units for both the real specimen and the image.

Magnification factors and real dimensions

Using magnification factors to find real dimensions

Magnification factors on images can be used to calculate the actual size of features shown on drawings and photographs of biological specimens. For example, in a photomicrograph of a cell individual features can be measured if the magnification is stated.

WORKED EXAMPLE

The magnification factor for the image of the open stoma is ×5000. This can be used to find out the actual size of any part of the cell. For example, how long is one guard cell from A to B?

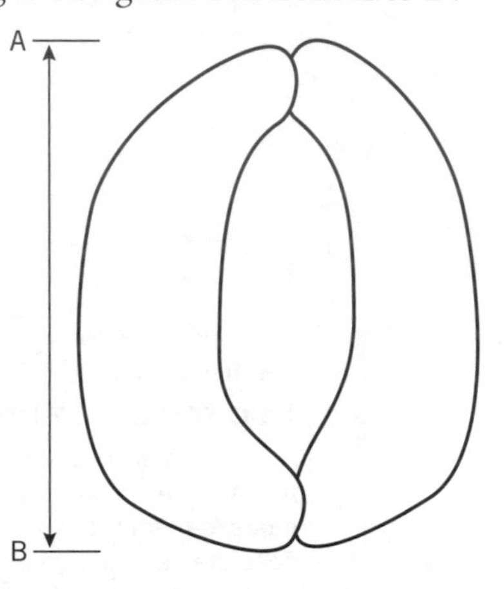

First measure the length of the guard cell as precisely as possible. In this example the image of the guard cell is 52 mm long.

Next convert this measurement to units appropriate to the image, in this case you should use μm because it is a cell.

So the magnified image is $52 \times 1000 = 52\,000$ μm

Remember size of image ÷ real size = magnification. Rearrange this equation to get:

Size of image ÷ magnification = real size

So the real length of the guard cell is now found by dividing:

$52\,000 \div 5000 = 10.4$ μm

SUMMARY QUESTION

1 For each of the images shown, use the magnification factor to determine the actual size of the object indicated.

a The length of the head and body of a spider

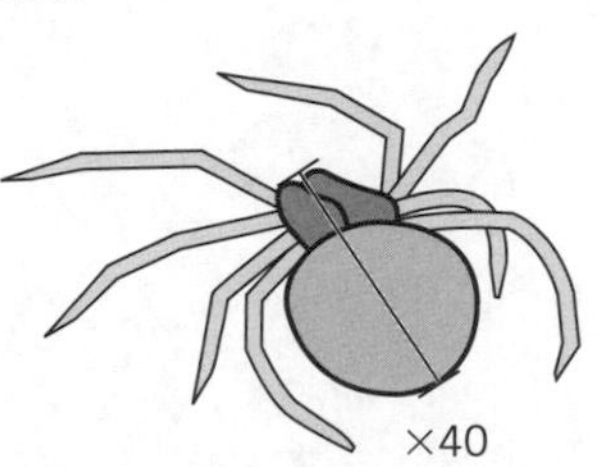

b The width of a Golgi apparatus

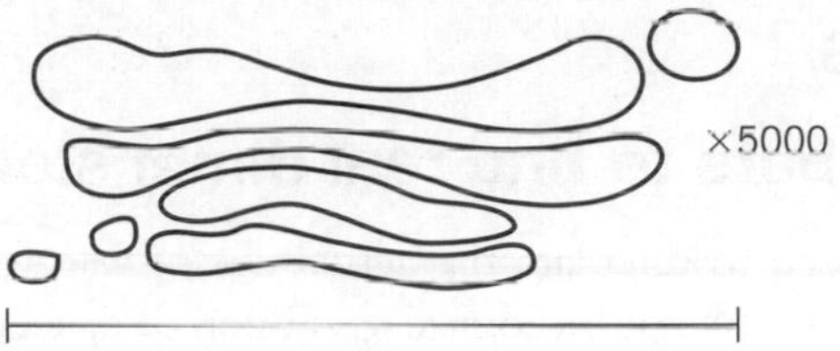

×5000

c The length of a bacterial cell

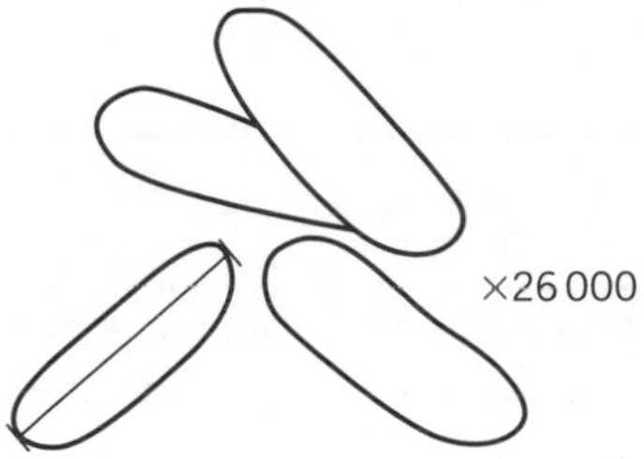

×26 000

d The length of the antenna of a butterfly

×0.7

e The length of a fern sporangium

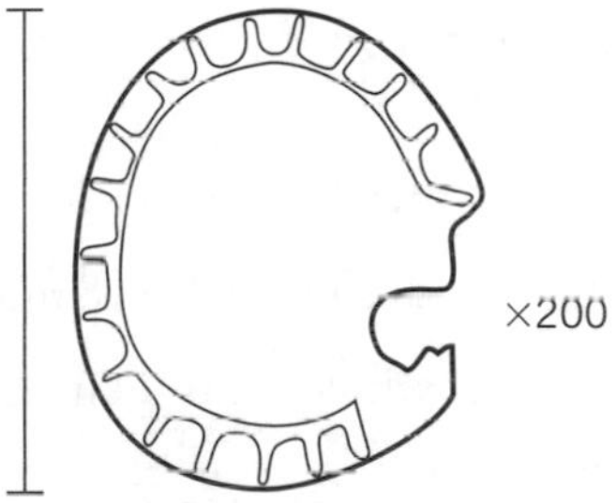

×200

REMEMBER:
size of an image ÷ real size = magnification

Rearrange this equation to find any of the three values.

Use an equation triangle to help:

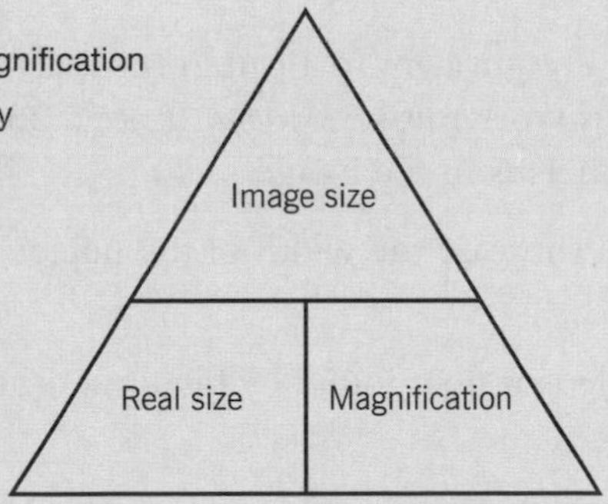

Image size

Real size | Magnification

Scale bars

Using scale bars to find real dimensions

Scale bars can be used to calculate the actual size of features shown on drawings and photographs of biological specimens. For example, in a photomicrograph of a cell individual features can be measured by using the scale bar.

WORKED EXAMPLE

In the image of a frog leucocyte, the scale bar at the bottom represents 1.5 μm. This can be used to find out the actual size of any part of the cell.

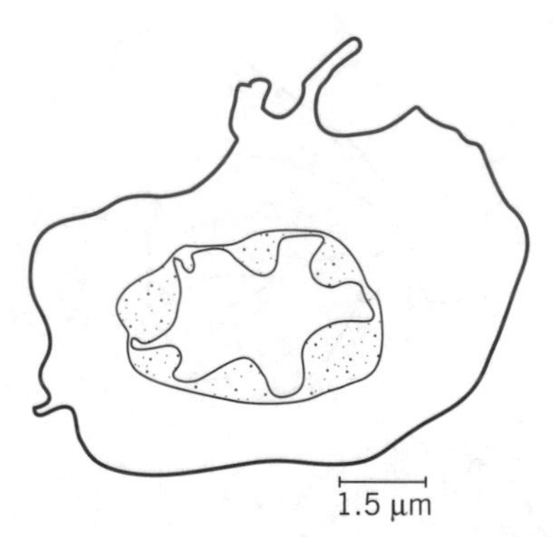

For example, what is the maximum diameter of the nucleus?

Step 1: measure the scale bar, which in this example is 8 mm long.

Step 2: if the scale bar is not already in whole units then work out the equivalent of 1 μm on the image.

- If 1.5 μm is equivalent to 8 mm then 1 μm is equivalent to $8 \div 1.5 = 5.33$ mm
- Thus every 5.33 mm on the image is equivalent to 1 μm in the real specimen. All you need to do now is work out how many times 5.33 mm divides into the dimension of the nucleus in the image.

Step 3: measure the desired feature, in this case the width of the nucleus. The width of the nucleus is 31 mm.

Step 4: the real dimension of the nucleus is now found by dividing using the sum:

length of the feature on the image ÷ length of 1 μm on the image

$31 \div 5.33 = 5.8$ μm

SUMMARY QUESTION

1 For each of the images shown, use the scale bar to determine the actual size of the object indicated.

a Length of the *Euglena*

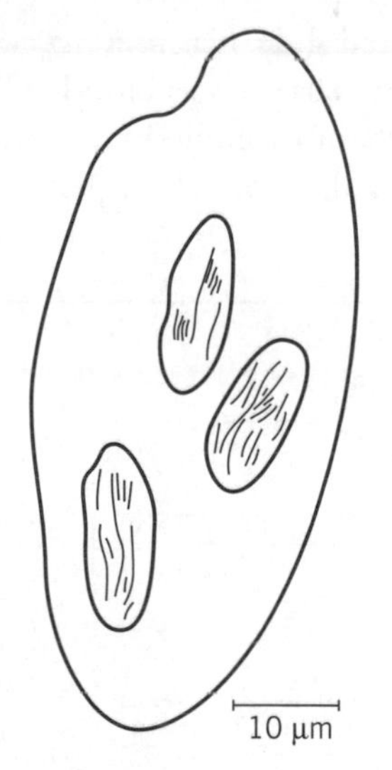

b Width of the widest xylem cell

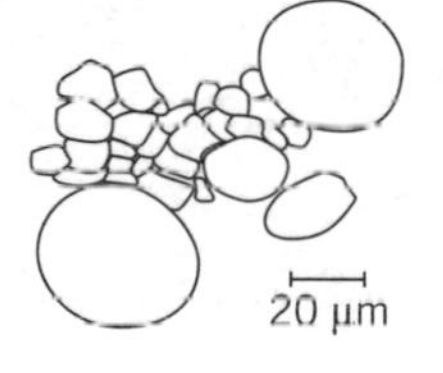

c Width of the insect's compound eye

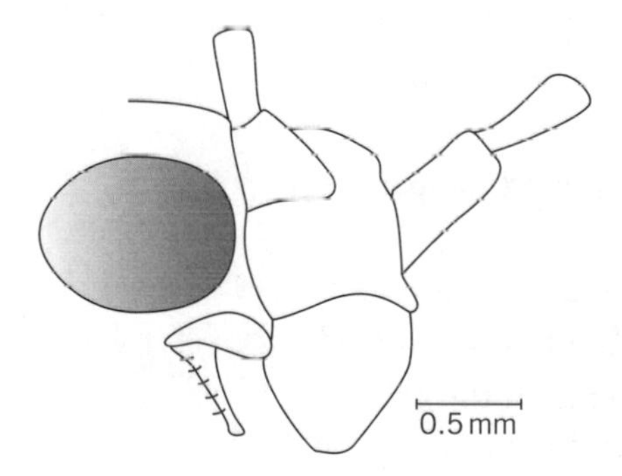

d Mean length of the yeast cells

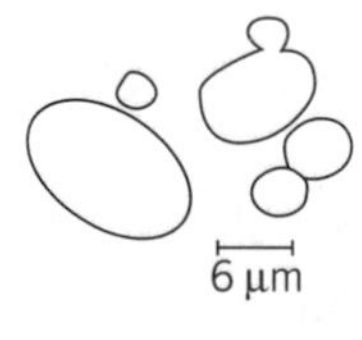

REMEMBER: Take care when measuring the scale bar and don't round up numbers until the end of the calculation.

Haemocytometers

Using a haemocytometer

A haemocytometer is a specially constructed slide. It is used to count the cells of bacteria or yeasts cultures or other cells such as blood cells. The slide is designed with a special counting chamber etched with precisely marked squares. The slide accommodates a drop of cell suspension exactly 0.1 mm deep.

WORKED EXAMPLE

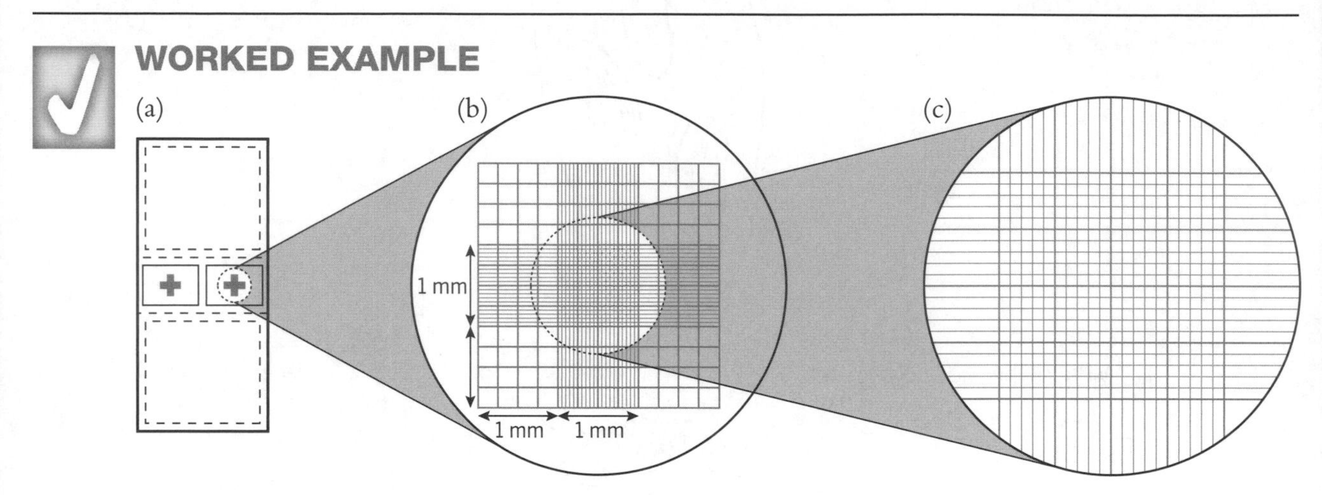

The counting area is divided by etched lines in such a way that they form small squares in the central region.

How many cells are in a sample of fresh blood?
Note: the cell suspension must be stirred so the cells are uniformly distributed before putting a sample on the haemocytometer. Often the sample is diluted so that it is feasible to count all the cells easily.

- A sample of fresh blood is diluted to 1% using isotonic saline (one part stirred blood to 99 parts saline).
- A drop of the blood is placed on the haemocytometer and the cells visible in eighty small squares are counted. The number of cells is found to be 880.
- Each small square is exactly 0.05 mm wide. Since the depth of the cell suspension is precisely 0.1 mm the volume of cell suspension visible above one small square is $0.05 \times 0.05 \times 0.1 = 2.5 \times 10^{-4}$ or $0.000\,25\ \text{mm}^3$.
- So the volume of blood above 80 squares is equivalent to $80 \times 0.000\,25\ \text{mm}^3 = 0.02\ \text{mm}^3$
- $0.02\ \text{mm}^3$ is $\frac{1}{50}$th of $1\ \text{mm}^3$ and this contains 880 cells.
- So the number of cells per whole mm^3 of diluted blood $= 50 \times 880 = 44\,000$.
- The diluted blood is 1% or $\frac{1}{100}$ of normal undiluted blood.
- So multiply by 100 to find the number of cells in undiluted blood.
- Which is $44\,000 \times 100 = 4\,400\,000$ per mm^3.

REMEMBER: Accurate counts depend on observing many small squares and using a stirred cell suspension. Remember to take dilution into account when calculating the final totals.

SUMMARY QUESTIONS

1 A sample of bacteria was diluted with an equal volume of sterile broth. When counted the average number of cells per small square was 17. What was the population of cells in the original culture?

2 An undiluted human cell culture was counted and there were 5 cells per small square.

  a How many cells were in each cm^3 of the original culture?

  b Assuming the number of cells doubles every 24 hours, how many hours would have passed when the number of cells per small square reached 23? Plot a log graph of the data and use it to estimate the answer. (Hint: you will need to convert 23 to its log before reading the time off the graph.)

3 A culture of bacteria needs to be at a density of at least 12×10^6 cells per mm^3 before a specific metabolite can be detected. If the culture is diluted to 2%, how many cells would be in each small square of a haemocytometer when the culture first reached this population density?

Water potential

Understanding water potential

Water potential (Ψ) is a measure of the tendency for water molecules to diffuse into or out of solutions and cells by osmosis. The key variables affecting water potential are the contribution made to the solution by the solutes (solute potential Ψ_s) and the pressure of the plasma membrane or cell wall pushing on the contents of the cell (pressure potential Ψ_p). Knowing the water potential allows predictions to be made about the direction of water diffusion between cells and between cells and solutions.

Cells in solutions

Unlike animal cells, when water diffuses into plant cells they do not burst because they have a cellulose cell wall. Pressure potential (Ψ_p) builds up and prevents more water entering. When the water potential of the plant cell is equal to the water potential of the solution surrounding it, there will be not net movement of water molecules (i.e. equilibrium is reached).

WORKED EXAMPLE

Water potential, solute potential and pressure potential have the following relationship:

$$\Psi = \Psi_s + \Psi_p$$

Ψ has a maximum value of zero (distilled water) and becomes negative as solute is dissolved.

Ψ_s is always a negative value, which becomes more negative as more solute is dissolved.

Ψ_p is a positive pressure and is exerted by either the plasma membrane or the cell wall.

The units are kPa.

a) The diagram shows two adjacent plant cells. Calculate the value of Ψ_{cell} for each one and determine the direction of any osmotic diffusion.

Cell A	Cell B
$\Psi_s = -700$ kPa	$\Psi_s = -500$ kPa
$\Psi_p = 200$ kPa	$\Psi_p = 100$ kPa

Cell A	Cell B
$\Psi_{cell} = \Psi_s + \Psi_p$	$\Psi_{cell} = \Psi_s + \Psi_p$
$\Psi_{cell} = -700 + 200$ kPa	$\Psi_{cell} = -500 + 100$ kPa
$\Psi_{cell} = -500$ kPa	$\Psi_{cell} = -400$ kPa

Cell B has a higher water potential (nearer to zero) so water diffuses from cell B to cell A by osmosis.

b) What will be the water potential of the cells at equilibrium?

To answer this, remember that at equilibrium Ψ of the two cells must be equal. So add the two values of Ψ then divide by two.

In this example $(-500 + -400) \div 2 = -450$ kPa

c) If cell A was placed in a bathing solution of $\Psi = -200$ kPa what would be the direction of osmotic diffusion and the value of Ψ_{cell} and Ψ_p at equilibrium?

The solution has a higher water potential so water diffuses from the solution into cell A by osmosis, raising the pressure potential. This causes the cell's water potential to increase until it is exactly equal to that of the solution and no further osmotic influx is possible.

REMEMBER: Ψ has a maximum value of zero. The more negative Ψ, the lower the water potential. Water diffuses from high to low water potential.

NOTE: This is a hypothetical situation: the volumes of water actually moving are so tiny that they have a negligible effect on Ψs, which is therefore assumed to remain unchanged.

At equilibrium the water potential of the cell will be equal to that of the solution, −200 kPa.

Therefore $\Psi_{cell} = \Psi_s + \Psi_p$ now gives $-200 = -700 + \Psi_p$

So Ψ_p must be 500 kPa.

> **NOTE:** Incipient plasmolysis is the point at which a plant cell placed in a solution with a lower water potential is just at the point of achieving plasmolysis. At this point $\Psi_p = 0$ and any further reduction in the water potential of the bathing solution will result in plasmolysis occurring, which will be visible with a microscope.

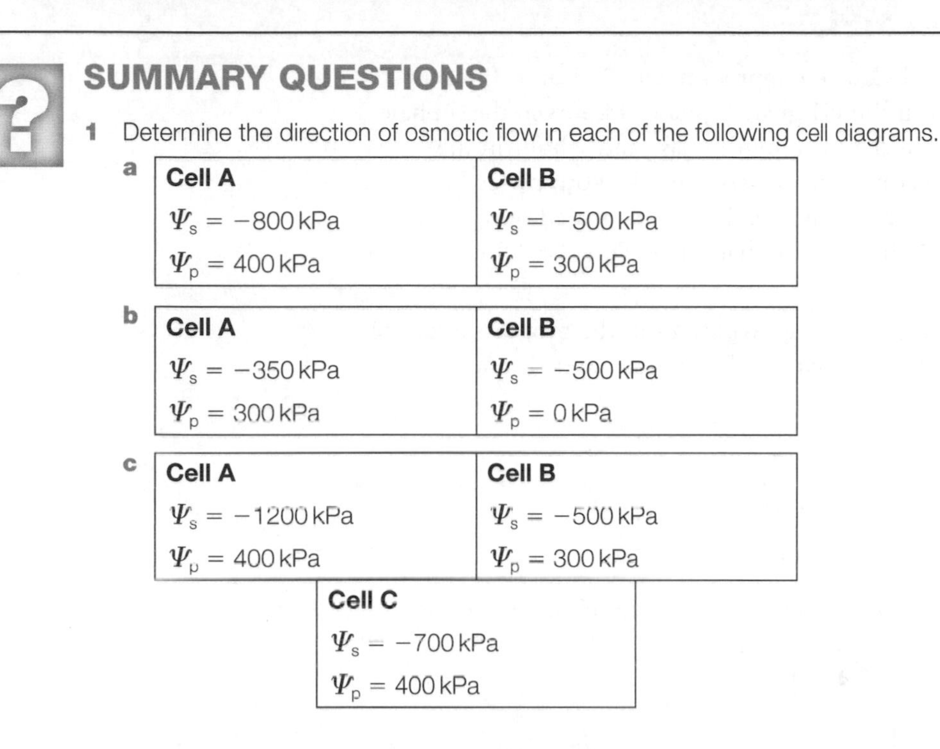

SUMMARY QUESTIONS

1 Determine the direction of osmotic flow in each of the following cell diagrams.

a

Cell A	Cell B
$\Psi_s = -800$ kPa	$\Psi_s = -500$ kPa
$\Psi_p = 400$ kPa	$\Psi_p = 300$ kPa

b

Cell A	Cell B
$\Psi_s = -350$ kPa	$\Psi_s = -500$ kPa
$\Psi_p = 300$ kPa	$\Psi_p = 0$ kPa

c

Cell A	Cell B
$\Psi_s = -1200$ kPa	$\Psi_s = -500$ kPa
$\Psi_p = 400$ kPa	$\Psi_p = 300$ kPa

Cell C
$\Psi_s = -700$ kPa
$\Psi_p = 400$ kPa

2 The table gives the initial values of Ψ_s and Ψ_p of some plant cells when placed into bathing solutions of given solute potential.
For each cell, calculate:

a the initial Ψ_{cell}

b the direction of osmotic flow

c the values of Ψ_{cell} and Ψ_p at equilibrium.

Cell	Initial value of Ψ_s/kPa	Initial value of Ψ_p/kPa	Ψ_s of the bathing solution/kPa
A	−1400	700	0
B	−800	0	0
C	−1100	400	−500
D	−900	600	−400

3 A plant cell with $\Psi_s = -600$ kPa and $\Psi_p = 200$ kPa is placed into a bathing solution with $\Psi_s = -700$ kPa. What is the initial value of Ψ_{cell}?
In what direction will osmotic flow occur?

4 A cell at incipient plasmolysis with a $\Psi_s = -2000$ kPa is placed into a bathing solution with a $\Psi = -800$ kPa. Determine the direction of osmotic flow. Calculate the values of Ψ_{cell} and Ψ_p for the cell when it reaches equilibrium.

Pie charts: the cell cycle

Converting numbers to pie charts: the cell cycle

Sometimes it is useful to present data in the form of a pie chart. This is often the case where a fixed period of time or total quantity is subdivided into parts. The parts can then be displayed easily using a pie chart. One example of data that can be presented on a pie chart is the cell cycle.

WORKED EXAMPLE

The human cell cycle lasts for approximately 24 hours. Typically 11 hours are spent in the G1 growth phase, 8 hours in the S phase (DNA replication), 4 hours in the G2 growth phase and 1 hour in the M (mitotic) phase. The duration of the phases is worked out by feeding marker chemicals to the cells. For example, thymidine is used only in the S phase, so it is used to mark the onset of that phase. Time for DNA mass to double represents the length of S.

To represent this on a pie chart begin by working out the proportion of the total circle to be allocated to each segment. The whole circle is 360°.

The segment for G1 is $\frac{11}{24} \times 360 = 165°$

The segment for S is $\frac{8}{24} \times 360 = 120°$

The segment for G2 is $\frac{4}{24} \times 360 = 60°$

The segment for M is $\frac{1}{24} \times 360 = 15°$

Draw a circle to represent the perimeter of the chart. Add one radius line.

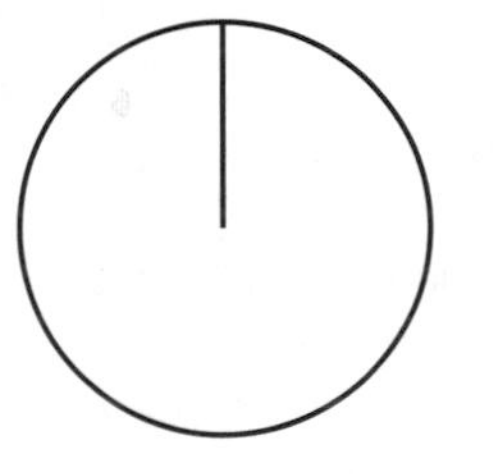

Next use a protractor to mark 165° from this and draw a second radius line.

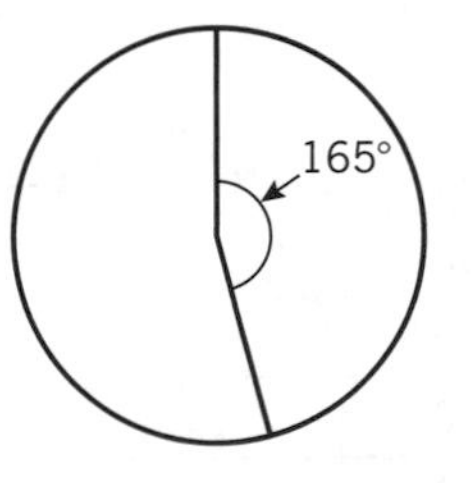

Mark the remaining segments in the same way.

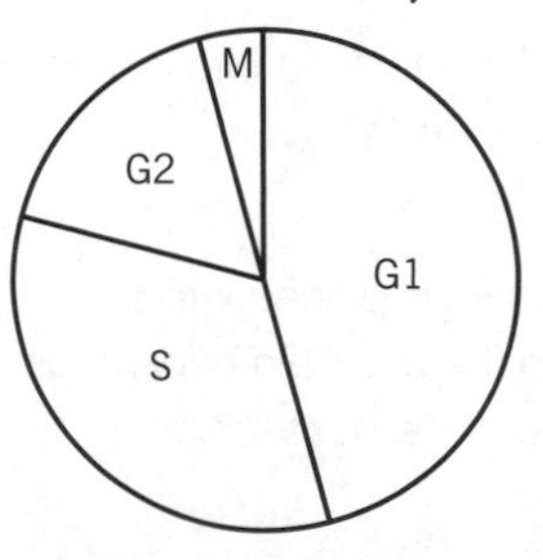

REMEMBER: Label the segments using a suitable legend. Colour can be helpful in picking out the segments.

SUMMARY QUESTIONS

1 Represent the following cancer cell cycle in a pie chart.

G1 = 5 hours
S = 7 hours
G2 = 3 hours
M = 1 hour

2 In a 24 hour cell cycle the segment of the chart representing the S phase was 102°. How long was the phase?

3 In rapidly dividing embryo cells the G1 and G2 phases may be omitted. The cycle can be reduced to 30 minutes duration. If the ratio between S and M phases remains constant at 8 : 1, what angle should define the segment for S phase?

4 In a certain cell line the M phase lasts for 75 minutes. If the cells double in number every 18 hours, what is the angle of the segment defining the whole interphase?

5 In a cell assay the DNA mass per cell doubled in 7 hours. Thymidine was first metabolised by the cell 12 hours after mitosis had been completed. The cell cycle was 28 hours.

a If the M phase was 1 hour, how long was the G2 phase?

b Draw a pie chart to represent this cell cycle.

Solutions at known concentrations

Making solutions at known concentrations

Experimental work in biology often requires the use of known dilutions of chemicals. For example, different salt or sugar solutions in osmosis experiments, different substrate or enzyme concentrations, varying bicarbonate ions in photosynthesis experiments.

WORKED EXAMPLE

There are two commonly used methods for making solutions at known concentrations.

Molar concentrations

The first is to make molar concentrations, for example specific sugar solutions for osmosis experiments.

First ascertain the molecular formula of the sugar, e.g. if sucrose is the sugar to be used the formula is $C_{12}H_{22}O_{11}$.

Next calculate the molecular mass by adding together the contributions of each constituent element. The atomic mass of carbon is 12, hydrogen is 1 and oxygen is 16.

So the molecular mass is $(12 \times 12) + (22 \times 1) + (11 \times 16) = 342$

If the molecular mass is expressed in grams then the quantity described is one mole of the substance.

To make a one molar solution, dissolve the molecular mass of the substance in one litre of distilled water. In this case that will be 342 g of sucrose in 1 dm^3 of water. The concentration is 1 mol dm^{-3}.

This ratio can be modified to give lesser volumes. For example, to make only 500 cm^3 of 1 mol dm^{-3} simply halve the quantities $\left(\frac{500}{1000}\right) \times 342 = 171$ g plus 500 cm^3 water; to make only 10 cm^3 of 1 mol dm^{-3} use $\left(\frac{500}{1000}\right) \times 342 = 3.42$ g plus 10 cm^3 water.

Percentage solutions

The second method is to make up percentage solutions. More complex molecules, such as enzymes, are often made up in this way. Percentage solutions are made by calculating the mass of the powdered or crystalline enzyme.

A 5% solution means 5 g out of every 100 g. Remember that distilled water has a mass of 1 g per cm^3, so a 5% solution is 5 g of the enzyme dissolved in 95 cm^3 of water.

> **REMEMBER:** Reagent bottles commonly list the molecular mass on the label. This can save you having to calculate this from scratch.
>
> A common error in making % solutions is always to dissolve in 100 cm^3. Remember the % is parts **out of** 100, not parts added to 100!

SUMMARY QUESTIONS

Work out the recipes for the following solutions:

1. 1 litre of 1 mol dm^{-3} sodium chloride
2. 25 cm^3 of 0.5 mol dm^{-3} glucose
3. 50 cm^3 of 0.1 mol dm^{-3} sodium bicarbonate
4. 500 cm^3 of 1% amylase
5. 65 cm^3 of 4% invertase
6. 750 cm^3 of 1% starch.

Serial dilutions and very small concentrations

Making serial dilutions and very small concentrations

Sometimes an experiment will require a series of different concentrations of the same chemical. Examples include a series of different molarity solutions in an osmosis experiment or dilutions of bacterial cultures during population count experiments. There are also occasions when the concentrations of solutions required are very small, so weighing introduces great inaccuracy. This is particularly true in some plant hormone experiments. In such cases serial dilution techniques can be used to make the required solutions.

WORKED EXAMPLE

To make a dilution series (i.e. a consecutive sequence of concentrations) you need to use the correct ratios.

Example 1: make a simple dilution series of 1 mol dm^{-3} sodium chloride for an osmosis experiment.

Volume of 1 mol dm^{-3} NaCl required/cm^3	Volume of distilled water required/cm^3	Final concentration of NaCl/mol dm^{-3}
100	0	1
90	10	0.9
80	20	0.8
45	55	0.45
etc.		

Begin by writing the recipe table as above. This will also help you to judge how much stock 1 mol dm^{-3} solution to make at the outset.

Example 2: make a true serial dilution (i.e. a repeated dilution) of a bacterial culture prior to dilution plate counting.

In this technique, each dilution made is used as the start point for the next one.

Mixture			Final dilution expressed as:		
			Fraction	Decimal	Standard form
A	1 part stock bacterial broth culture	9 parts sterile broth	$\frac{1}{10}$	0.1	10^{-1}
B	1 part of mixture A	9 parts sterile broth	$\frac{1}{100}$	0.01	10^{-2}
C	1 part of mixture B	9 parts sterile broth	$\frac{1}{1000}$	0.001	10^{-3}
etc.					

Remember to make up sufficient volume of sterile broth to use for the subsequent dilution steps.

Example 3: use serial dilution to make very small dilutions accurately, such as plant hormone solutions for tissue culture.

If the objective is to make a solution of 0.2 g dm^{-3} of kinetin hormone, a first glance would suggest a recipe dissolving 0.2 g in a litre of water. However, this is very wasteful as far too much solution is made. It is also expensive; one gram of kinetin costs about £50! A real experiment might actually require only 50 cm^3 of the 0.2 g dm^{-3} solution. This would require adding $\frac{50}{1000} \times 0.2 = 0.01$ g of the kinetin, but weighing tiny quantities can lead to inaccuracy.

Instead, begin by making a stock solution by dissolving an easily weighed mass of the kinetin in a small volume. If the solution is made with 0.1 g kinetin in 10 cm^3 solvent the stock solution has a concentration of 10 g dm^{-3}. So every 1 cm^3 of this solution will contain 0.01 g of kinetin. Adding 1 cm^3 of the stock solution to 49 cm^3 of solvent produces the desired 50 cm^3 of solution at concentration 0.2 g dm^{-3}. Hormones can often be bought ready made as stock solutions to make life easier.

REMEMBER: By using repeated dilutions of existing solutions it is easy to make different concentrations accurately without waste.

SUMMARY QUESTIONS

Work out the recipes for the following solutions:

1. A dilution series producing 10 cm^3 each of 0.1, 0.3, 0.5, 0.7 and 0.9 mol dm^{-3} sucrose solutions.
2. A serial dilution of a plant mineral solution to give 500 cm^3 each of dilutions at 10^{-1}, 10^{-2}, 10^{-3}, 10^{-4}, and 10^{-5}.
3. What is the final concentration of auxin if 0.2 g is dissolved in 10 cm^3 of solvent and then 0.5 cm^3 of this stock solution is diluted with a further 199.5 cm^3 of growth medium?
4. A broth culture of bacteria contains 10^6 cells per mm^3. Design a dilution series, using no more than 100 cm^3 of sterile broth, that would contain a single cell per mm^3.
5. Sometimes the initial solvent for plant hormones is alcohol. This will evaporate from the subsequent solution, so correct the volumes to take this into account. Suggest an economical recipe to make 100 cm^3 of 0.0001 g dm^{-3} cytokinin.

Calorimetry and energy values of foods

Using calorimeters to calculate energy values of foods

In living cells, respiration releases the energy from food molecules using a sequence of enzyme controlled reactions. In a laboratory a sample of food may be burned completely in an oxygen atmosphere so that the energy is released quickly as heat. The equipment used is called a food or 'bomb' calorimeter. All the heat energy released is absorbed by a standard volume of distilled water. The temperature rise of the water can then be used to calculate the total energy that was released from the burning food.

Distilled water must be used as solutes change the specific heat capacity. 1 cm^3 of distilled water has a mass of 1 g.

WORKED EXAMPLE

Energy value in a peanut

Calculations rely on the use of a constant called the specific heat capacity. This is defined as the heat energy required to heat a unit mass of a substance by one degree Celsius. For distilled water the actual specific heat capacity is 4.186 joules per gram °Celsius but this figure is commonly quoted as $4.2\,J\,g^{-1}\,°C$.

Neil burned a sample of peanuts and recorded the following measurements:

Initial mass of food sample/g	0.9
Mass of ash residue after burning/g	0.1
Mass of water heated/g	1000
Temperature of the water at the start/°C	18
Final temperature of the water/°C	23

The energy released by burning the nut is

$4.2 \times$ the mass of water (g) $\times$ the temperature rise (°C) joules.

First calculate the temperature rise of the water:

Final temperature $-$ initial temperature °C

$= 23 - 18\,°C$

$= 5\,°C$

Now calculate the energy released:

$4.2 \times$ the mass of water (g) $\times$ the temperature rise (°C) joules

$4.2 \times 1000 \times 5 = 21\,000\,J$

The actual mass of peanut burned was $0.9 - 0.1 = 0.8\,g$

So the actual energy content of the peanut can now be calculated:

Energy released (J) $\div$ mass of food burned (g)

$= 21\,000 \div 0.8\,J\,g^{-1}$

$= 26\,250\,J\,g^{-1}$

To quote energy values in kJ divide by 1000.

$26\,250 \div 1000 = 26.25\,kJ\,g^{-1}$

Food packaging often quotes energy values as kJ per 100 g of the food.

Neil's calculated value would therefore be converted like this:

$26.25 \times 100 = 2625\,kJ\,(100\,g)^{-1}$

REMEMBER:

Use the right units

Food labelling sometimes uses calories or kilocalories as units. You should use joules (J) or kilojoules (kJ). It is possible to convert one unit to another, as 1 calorie is approximately 4.2 J.

SUMMARY QUESTIONS

1 1.2 g of muesli bar was burned in a food calorimeter. The temperature of the water at the start was 18.5 °C and after burning was complete the new temperature was 22.1 °C. There was 0.1 g of ash remaining. The calorimeter contained 1 litre of water. Calculate:

a the energy released by burning the food sample

b the energy content of the muesli bar in $kJ\,g^{-1}$.

2 Undertake the following conversions:

a 7.4 $kJ\,g^{-1}$ into $J\,g^{-1}$

b 1208 kJ $(100\,g)^{-1}$ into $J\,g^{-1}$

c 9.6 kJ $(0.6\,g)^{-1}$ into $kJ\,kg^{-1}$.

3 In a test a sample of food gave an energy release of 10.5 kJ when burned in a calorimeter containing 1 litre of water.

a If the final temperature of the water was 19.5 °C, what was the initial temperature of the water?

b When converted to $kJ\,g^{-1}$ the energy value of the food was 17.5 $kJ\,g^{-1}$. What mass of food must have been burned in the test?

4 0.3 g of rump steak was burned in a food calorimeter. The temperature of the water at the start was 16.0 °C and after burning was complete the new temperature was 17.9 °C. There was 0.05 g of ash remaining. The calorimeter contained 500 cm^3 of water. Calculate:

a the energy released by burning the food sample

b the energy content of the steak in $kJ\,g^{-1}$.

Chromatography and R_f values

Working with chromatography and R_f values

Compounds of different molecular weights are often separated by the technique of solvent chromatography, for example separation of photosynthetic pigments or the separation of amino acids. R_f values can be used to identify the compounds.

WORKED EXAMPLE

After separation, the molecules of the compounds show up as spots along the chromatography paper. R_f values are ratios of the distance to the front of the spots to the total distance moved by the solvent, hence R_f or 'relative front' value.

To find R_f for a given spot use the formula:

$$R_f = \frac{\text{distance moved by the compound}}{\text{distance moved by the solvent}}$$

The diagram shows an example in which the solvent has moved 100 mm. The front of the spot from compound *a* has reached 60 mm.

So the R_f value for compound *a* is $60 \div 100 = 0.6$

Using the same method, compound *b* has an R_f value of $20 \div 100 = 0.2$

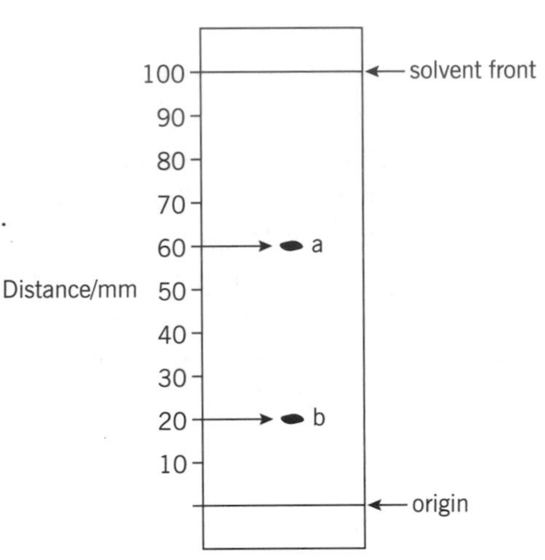

REMEMBER: R_f values are ratios. The distances travelled vary between different solvents, so R_f values can only be compared if the same solvent is used.

SUMMARY QUESTIONS

1 Calculate the R_f values of the following measurements:

Compound	Distance moved by the compound/mm	Distance moved by solvent front/mm
a	144	200
b	38	100
c	36	150
d	27.5	50
e	30	100

2 Identify compounds **a–e** from Question **1** by using the reference table:

Amino acid	R_f value
alanine	0.38
arginine	0.20
asparagine	0.50
aspartic acid	0.24
cysteine	0.40
glutamine	0.13
glutamic acid	0.30
glycine	0.26
histidine	0.11
isoleucine	0.72
leucine	0.73
lysine	0.14
methionine	0.55
phenylalanine	0.68
proline	0.43
serine	0.27
threonine	0.35
tryptophan	0.66
tyrosine	0.45
valine	0.61

3 Find the missing distances in the table below.

	Distance moved by the compound/mm	Distance moved by solvent front/mm	R_f value
a	90		0.98
b		118	0.59
c		56	0.42
d	120		0.28

4 In Question **3** the four compounds are photosynthetic pigments described by a researcher called Carol Reiss in 1994. Use the internet to try and identify them.

Rates of reaction 1

Relative rates of reaction

When a biological experiment is monitored continuously at set time intervals the results can be displayed in a graph showing reaction progress, e.g. amount of product, against time. These graphs can easily be converted to rates graphs.

WORKED EXAMPLE

Concentration of protease/%	2.0	2.5	3.0	3.5	4.0
Time/s	Percentage transmission				
0	5	5	4	6	5
20	4	4	4	7	5
40	5	4	5	7	9
60	6	6	6	8	15
80	7	7	9	11	25
100	8	9	11	15	40
120	10	12	15	21	52
140	12	15	19	30	57
160	15	18	25	39	58
180	18	22	32	47	58
200	21	27	39	52	59
220	26	33	46	55	59
240	30	40	51	56	59
260	36	46	54	57	60

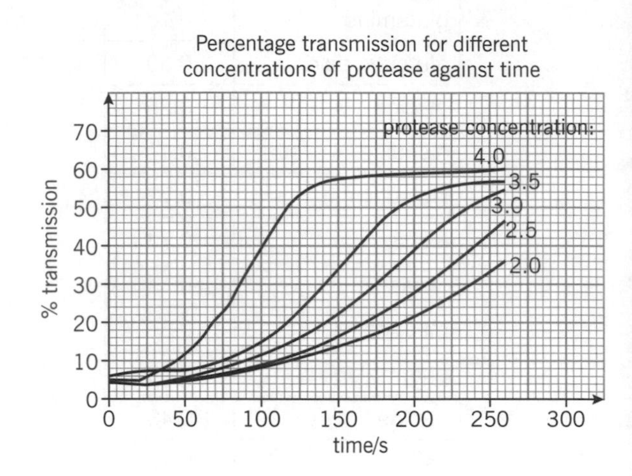

The example shows data collected using a colorimeter when following the digestion of a protein suspension by different concentrations of protease enzyme. As the digestion proceeds the solution clarifies and the percentage transmission rises.

In the time available only the reactions at 4% and 3.5% approach completion (the curve reaches a plateau).

A quick way to work out the rate of reaction is to select an end point that all curves reach, for example the time taken to reach 30% transmission. This is time t. Choose a point where the curve is rising steeply (i.e. beyond the phase of acceleration).

Convert the times to **relative rates** by calculating $\frac{1}{t}$. In this example the values of $\frac{1}{t}$ are very small, so multiply them by 1000 to get whole numbers as seen in the following table. The shorter the time taken to reach this point, the faster the reaction is proceeding so the larger the relative rate.

These can then be plotted to give a graph of relative reaction rate against concentration of the enzyme.

Concentration of protease/%	Time t taken to reach 30% transmission/s	$\frac{1}{t} \times 1000$
2.0	240	4.2
2.5	218	4.6
3.0	163	6.1
3.5	140	7.1
4.0	85	11.8

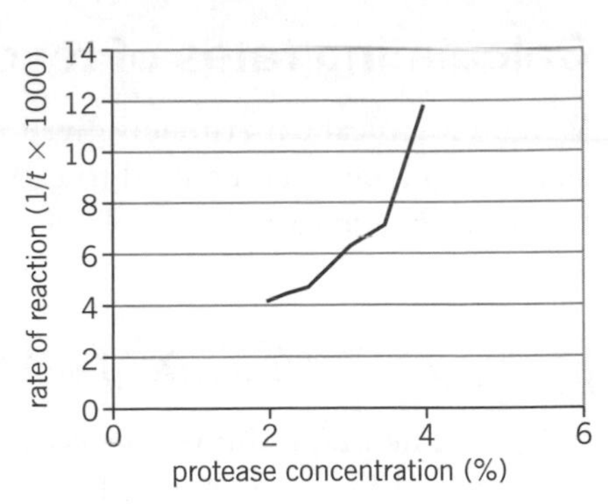

REMEMBER: The rates calculated in this way are relative. Remember to choose a sensible end point such as one on the steepest part of the curves.

SUMMARY QUESTIONS

1 **a** Plot the following data.

b Calculate relative rates of reaction for each temperature for the time taken to collect 20 cm^3 of gas.

c Plot a relative rates graph to show the effect of temperature on the rate of reaction.

Temperature/°C	Time/min									
	1	2	3	4	5	6	7	8	9	10
	Cumulative volume of gas produced over 10 min/cm^3									
10	0.5	1.0	2.0	2.5	4.0	6.0	9.0	12.5	16.5	21.0
20	2.0	3.5	6.0	9.0	13.0	18.5	24.0	29.0	33.0	37.0
30	4.0	8.0	13.0	19.5	27.0	33.5	38.0	42.0	45.0	45.0
40	10.0	19.0	28.0	35.0	41.5	47.0	49.0	51.5	54.0	54.0
50	20.0	42.0	51.0	56.0	59.0	61.0	62.0	63.0	63.0	63.0

2 In an experiment the time taken for a sample of milk to clarify when being digested by protease was measured. The data collected were as follows:

Enzyme concentration/%	Time taken to clarify the milk sample/s
2	460
4	213
6	135
8	102
10	79

a Calculate relative rates of reaction.

b Plot a suitable graph to show the effect of enzyme concentration on the rate of reaction.

Rates of reaction 2

Calculating rates of reaction

When biological experiments give trends in which the rate of reaction changes, tangents can be used to calculate the rate of reaction at any given point on the curve, for example, the fastest rate.

WORKED EXAMPLE

The graph shows data collected during the reaction between catalase and hydrogen peroxide. The volume of gas collected is plotted against time. What is the maximum reaction rate?

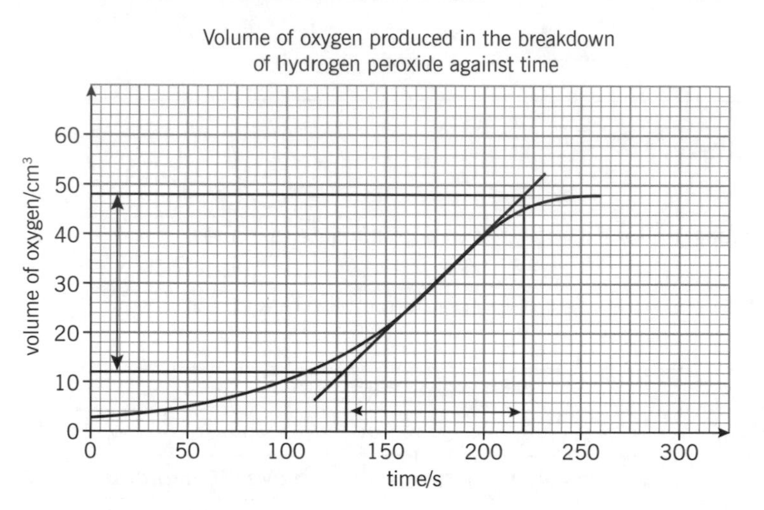

First construct a tangent line by drawing a line using a ruler. The tangent line should touch the curve at its steepest point.

Next use construction lines to find the values of x and y at any two selected points on the tangent, as shown.

Find the change in y and the change in x between the two selected points. In the example y goes from 12 to 48, a change of 36 cm^3 oxygen while x changes from 128 to 220 s, a change of 92 s.

Calculate the rate by dividing the change in y by the change in x $\left(\frac{dy}{dx}\right)$.

In the example $\frac{dy}{dx} = \frac{36}{92} = 0.39\ cm^3\,s^{-1}$.

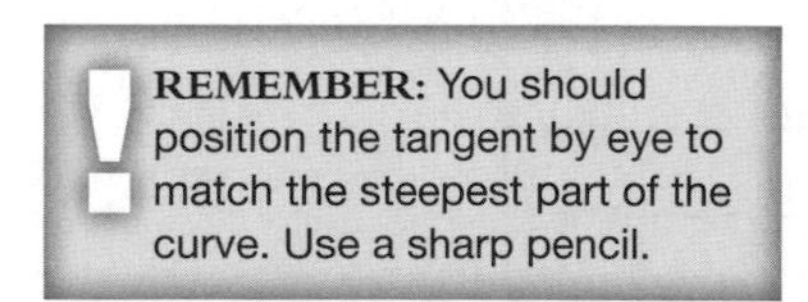

REMEMBER: You should position the tangent by eye to match the steepest part of the curve. Use a sharp pencil.

SUMMARY QUESTIONS

1 **a** Plot graphs of the following data.

b Draw suitable tangents on each curve.

c Use the tangents to calculate rates of reaction.

Concentration of bicarbonate/%	Time/s					
	30	60	90	120	150	180
	Cumulative volume of oxygen collected/mm^3					
0	0.5	1.5	2.5	4.5	7.0	10.0
1	1.0	2.0	3.5	6.5	10.0	15.0
2	2.5	6.0	11.0	17.0	24.0	35.5
3	6.0	14.0	27.0	41.5	50.5	52.0
4	10.0	25.0	42.5	51.0	52.5	53.0

2 The table shows data from an experiment in which glucose was being released by the digestion of starch.

a Plot the data on a suitable line graph.

b Use a tangent to calculate the maximum rate of reaction.

c Find the rate of reaction at:

i 8 minutes

ii 32 minutes.

Time/min	Glucose produced/ mmol dm^{-3}
5	2
10	6
15	12
20	22
25	25
30	28
35	29

Colorimeters

Using a colorimeter

A colorimeter is used to measure the amount of light penetrating a coloured solution as a measure of the density of the colour or turbidity of the solution being tested. Either transmission or absorbance of light may be measured. The colour of the test light is varied using filters, e.g. a red test light would be used to measure transmission by a blue solution; the deeper the blue colour the less red light penetrates. Examples might include the progress of an amylase-starch digestion by measuring the blue colour produced by testing with iodine solution or bacterial growth in broths by measuring turbidity.

WORKED EXAMPLE

Producing a calibration curve

Step 1: prepare a series of standard dilutions of starch solution. This is an opportunity to practise the methods described on pages 50 and 51.

Step 2: mix 3 cm^3 of the first starch solution with 1 cm^3 of a standard iodine solution in a cuvette (the special container used in a colorimeter) then measure the percentage absorbance of red light. Repeat this exactly with the remaining starch solutions.

Step 3: plot percentage absorbance against starch concentration and draw a smooth best fit line.

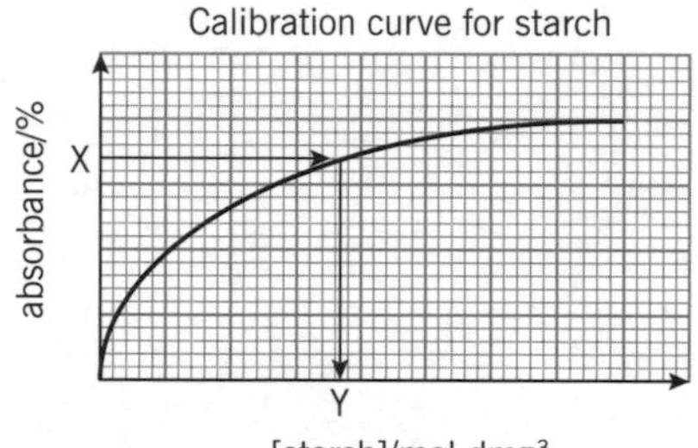

If an unknown starch solution is now tested the percentage absorbance X may be converted to concentration Y using the curve as shown. This method allows the concentration of a substrate to be monitored as a digestion reaction proceeds.

> **REMEMBER:** Make sure that volumes and concentrations of the iodine solution remain constant in all tests.
>
> Don't forget that this is iodine **solution**, not 'iodine', which is a solid violet crystal!

Producing an absorption spectrum

You can use a colorimeter to generate an absorption spectrum for a sample of chlorophyll extract. Prepare a suitable extract by using solvents in the same way as for chlorophyll chromatography experiments.

Set up the colorimeter to record percentage absorbance. Use a cuvette containing just the solvent as a blank to zero the instrument. Set the first filter in place and note the wavelength of light that passes. This is labelled on the filter. Record the percentage transmission. Repeat this measurement with each filter in turn. Your data will look like this:

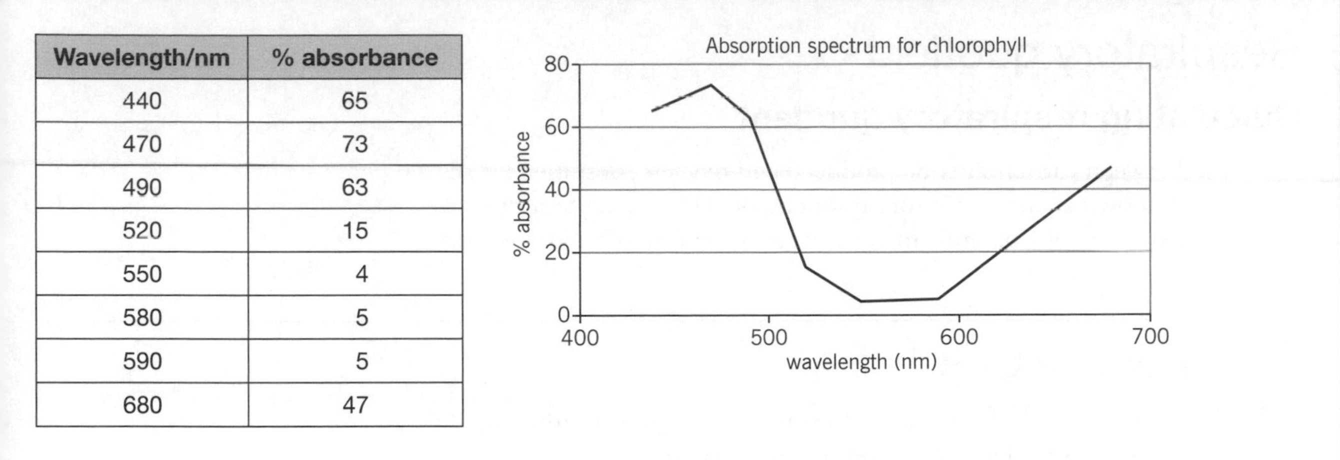

Wavelength/nm	% absorbance
440	65
470	73
490	63
520	15
550	4
580	5
590	5
680	47

SUMMARY QUESTIONS

In an experiment the absorbance of a number of standard starch solutions was measured.

Starch/%	5.0	4.5	4.0	3.5	3.0	2.5	2.0	1.5	1.0	0.5	0.1	0
Absorbance/%	99	96	91	86	81	75	67	58	45	27	6	0

1 Plot a calibration curve using the data in the table.

2 Find the concentration of starch remaining in a 5% starting solution when the absorbance was:

a 36%

b 50%

c 88%.

3 In an experiment on bacterial growth the turbidity of the culture was measured using a colorimeter recording percentage absorbance of blue light. A haemocytometer was used to calculate the number of cells present per mm^3 of the broth at each stage.

Cell count/millions per mm^3	0	1	2	3	4	5
Absorbance/%	0	9	22	33	44	55

Predict the population of cells that would produce absorbance of:

a 76%

b 18%

c 89%.

Respiratory quotient

Calculating respiratory quotient

The ratio between carbon dioxide produced and oxygen consumed by an organism is known as the respiratory quotient, or RQ. It is quite simple to calculate and gives valuable information about the nature of the respiratory substrate.

WORKED EXAMPLE

To find RQ it is necessary to measure the oxygen absorbed by the organism as well as the carbon dioxide evolved.

Assumption 1

In a simple case aerobic respiration will be
$C_6H_{12}O_6 + 6O_2 \rightarrow 6CO_2 + 6H_2O$ plus energy.

Therefore every mole of oxygen used is exactly replaced by a mole of carbon dioxide.

Assumption 2

If a substrate such as lipid or protein is being respired the organisms will be absorbing extra oxygen because they have to oxidise the lipid or protein at the start. More oxygen will be absorbed than carbon dioxide evolved.

Step 1:

Use the respirometer to record total oxygen uptake in the normal way. The distance moved by the meniscus in the capillary tube manometer can be calculated as long as the radius (*r*) of the tube is known. The distance moved by the meniscus is *h* and the volume indicated by this movement is found using the formula $\pi r^2 \times h$. This will be the total volume of oxygen absorbed and will include both respiration (O_2 *resp*)and oxidation (O_2 *oxid*). For example, if the meniscus moved 45 mm in a tube radius 0.5 mm, the volume absorbed would be $3.14 \times 0.5^2 \times 45 = 35.3\ mm^3$.

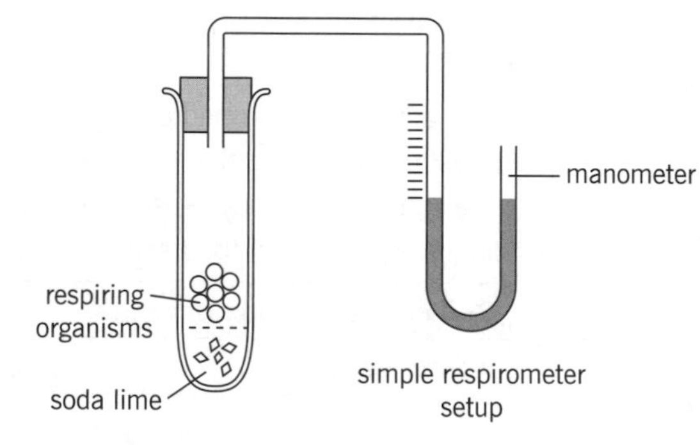

Step 2:

Remove the soda lime carbon dioxide absorber and measure any movement of the meniscus, again using the formula $\pi r^2 \times h$ to convert this to a volume. For example, if the meniscus now moves 4 mm, the volume of oxygen absorbed is $3.14 \times 0.5^2 \times 4 = 3.2\ mm^3$.

Since every oxygen molecule used in respiration is replaced by a molecule of carbon dioxide, the volume measured in the absence of the soda lime is the extra oxygen being absorbed for oxidation. This is the volume O_2 *oxid*.

Step 3:

Calculate the volume of carbon dioxide evolved
(remember O_2 *resp* = CO_2 evolved) using the formula below.

volume absorbed with soda lime present (O_2 *resp* + O_2 *oxid*) − volume absorbed with no soda lime (O_2 *oxid*)

$$35.3 - 3.2\ mm^3 = 32.1\ mm^3$$

Step 4:

Calculate the RQ using the formula $RQ = \frac{\text{volume of carbon dioxide evolved}}{\text{total volume of oxygen absorbed}}$

In this example $RQ = \frac{32.1}{35.3} = 0.91$

Interpretation:

Respiratory substrate	RQ
Carbohydrate	1.0
Lipid	0.7
Protein	0.8–0.9

REMEMBER: The RQ may be due to a combination of substrates.
Use larger masses of organisms in the respirometer to give clear meniscus movements to eliminate the proportional error effects of measuring very small distances.

SUMMARY QUESTIONS

1 Work out the missing values in the table below.

	Organism in the respirometer	Volume of oxygen absorbed with soda lime present/mm^3	Volume of oxygen absorbed with no soda lime present/mm^3	Volume of CO_2 evolved/mm^3	RQ	Interpretation
a	Peas	46		37		Protein stores being respired
b	Woodlice	39	0			
c	Starved woodlice	41	12			
d	Sunflower seeds		22	54		Lipid stores being respired

2 Work out the volumes of oxygen that have been absorbed when the meniscus moves the following distances in a 1 mm diameter capillary tube:

a 27 mm

b 78 mm

c 1 mm.

3 Suggest the RQ expected with fermenting yeast.

Spirometer data

Understanding spirometer data

A trace from a spirometer typically shows a descending trend such as that in the example below. This is because the peaks correspond to the height of the spirometer float chamber, which gets less with every breath as oxygen absorbed by the user is removed from below it (the exhaled carbon dioxide is removed by the soda lime in the filter). The trace is drawn on a graph with a known time base (x-axis) and calibrated volume axis (y-axis).

WORKED EXAMPLE

Measuring the tidal volume

This is the volume of one normal breath. First project construction lines to the y-axis. Then measure the volume using the scale on the axis (*tv* on the diagram).

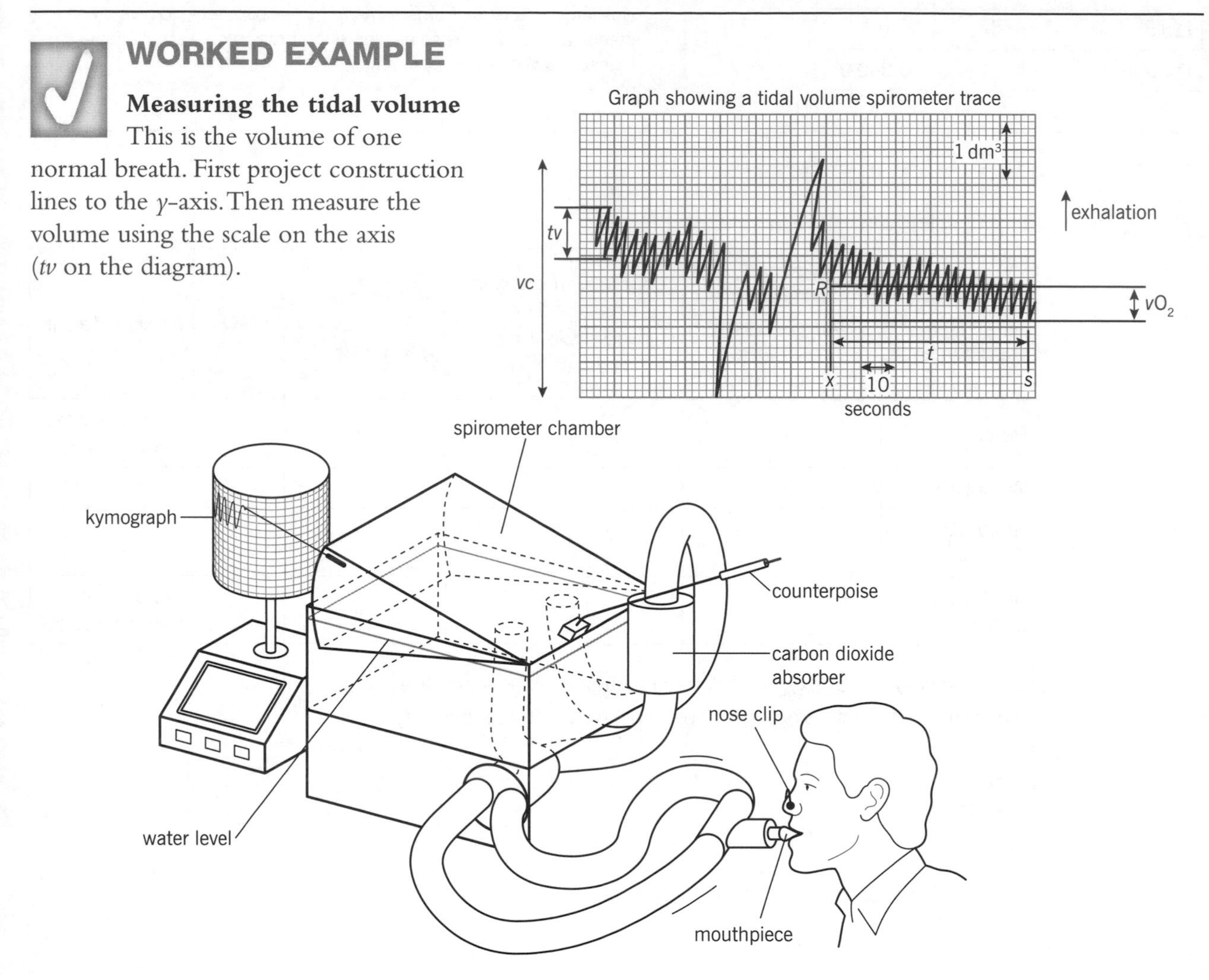

Measuring the vital capacity

This is the volume of the largest breath possible. First project construction lines to the y-axis. Then measure the volume using the scale on the axis (*vc* on the graph).

Measuring breathing rate

To do this, simply count the number of breaths within the span of one minute. In the example above $t = 60$ s and there were 21 breaths so the breathing rate is 21 breaths per minute.

Measuring the ventilation rate

This is the volume of air circulated through the lungs per unit time. Simply multiply the tidal volume by the breathing rate in breaths per minute. In the example above, during the span of time t 21 breaths $\times$ 0.7 dm^3 = 14.7 dm^3 min^{-1}.

STRETCH YOURSELF

You can make more advanced measurements from the trace.

Measuring the oxygen consumption per unit time

Begin by drawing a straight line joining the tops or bottoms of a range of breaths. This must be judged by eye, taking into account the natural variation in breaths in such traces.

Next use construction lines from any two chosen points to measure the volume of oxygen used (vO_2 on the graph) and the time taken (t on the graph).

Finally divide $\frac{vO_2}{t}$ to give a rate of absorption.

> **REMEMBER:**
> When making measurements use the same two points when projecting to the *x*- and *y*-axes. Convert units so that the calculated numbers make sense.

Example

In the trace above vO_2 is 0.6 dm^3 and $t = 60$ s

The rate of oxygen absorption is therefore $0.6 \div 60\,dm^3\,s^{-1} = 0.01\,dm^3\,s^{-1}$

It is useful to convert dm^3 to cm^3 to get rid of the small number.

$1\,dm^3 = 1000\,cm^3$ so to convert the units multiply by 1000, i.e. $0.01 \times 1000 = 10\,cm^3\,s^{-1}$

Measuring the oxygen absorbed by one breath

To do this count the number of breaths (i.e. the number of peaks) spanning the time t marked on the graph and divide the volume of oxygen absorbed by this number.

In the example above, t spans 21 breaths and the O_2 absorbed was 0.6 dm^3

So the volume of oxygen absorbed by each breath is $\frac{0.6}{21}\,dm^3$ per breath $= 0.03\,dm^3$ per breath

Again, convert this number by multiplying to get $0.03 \times 1000 = 30\,cm^3$ absorbed per breath.

SUMMARY QUESTIONS

1 For trace **a**, calculate:

- **a** the tidal volume
- **b** the breathing rate
- **c** the ventilation rate
- **d** the oxygen consumption per minute.

2 For trace **b**, calculate:

- **a** the ventilation rate
- **b** the vital capacity
- **c** the total oxygen consumption during the first 10 seconds.

3 For trace **c** calculate:

- **a** the initial breathing rate
- **b** the final breathing rate
- **c** the percentage change in ventilation rate.

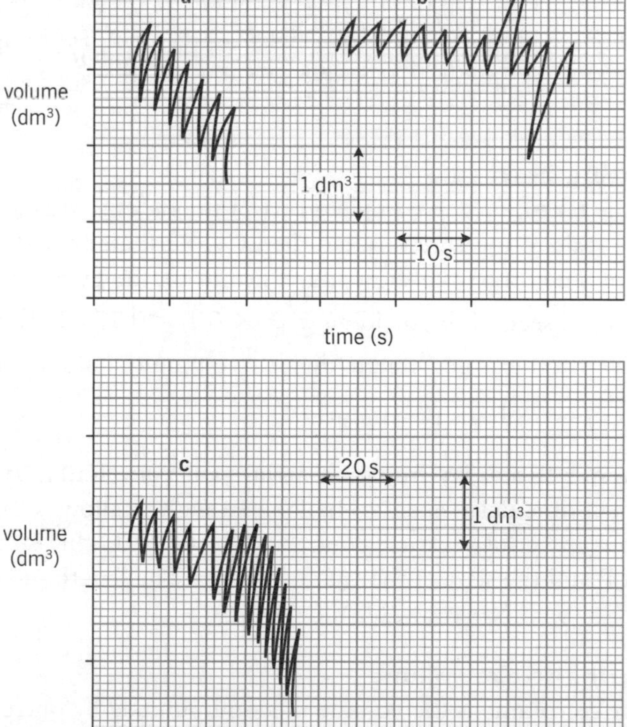

Electrocardiograms

Understanding an electrocardiogram

The electrical impulses transmitted through the heart during the heartbeat can be detected using electrodes on the surface of the chest and displayed on a monitor as a visible trace called an electrocardiogram. This, combined with pressure and volume records, can be used to measure quantities such as the heart rate, stroke volume and cardiac output.

WORKED EXAMPLE

The trace in the diagram illustrates a recording in which two complete beats of the heart appear. Pressure, volume and ECG are all recorded on the same time scale. The letters on the ECG follow an international convention and are used to identify specific parts of the sequence. P is the spread of impulses from the sinoatrial node to the atrial muscles causing them to contract. QRS is the spread of impulses through the ventricles causing them to contract. T is repolarisation of the ventricles.

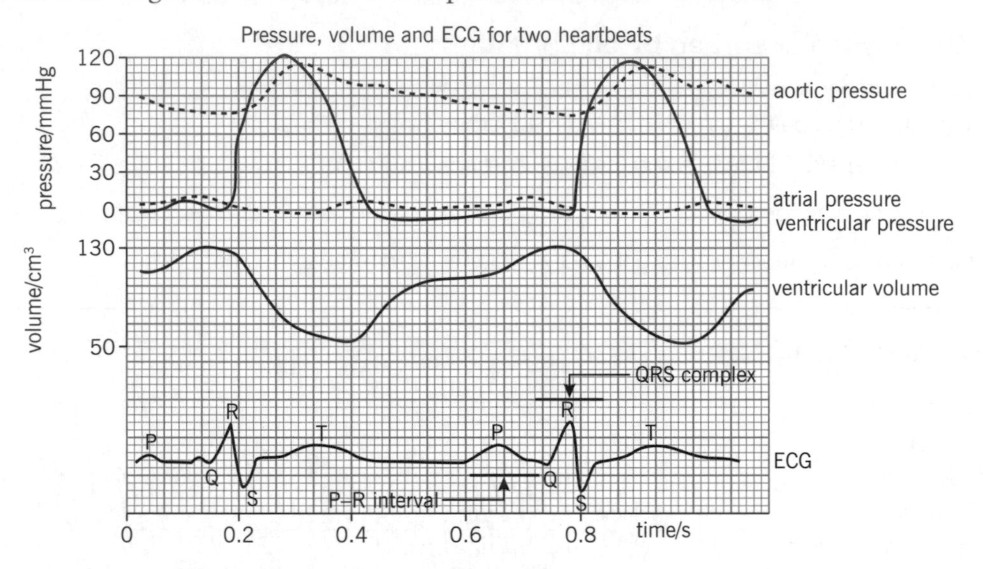

Measuring heart rate

The time taken for one complete beat of the heart is the distance between the R peaks. Use the scale to measure this. On the ECG illustrated the distance between the R peaks is 57.5 mm. The time scale shows that 7.5 mm on the graph = 0.1 s. Convert 57.5 mm on the graph to seconds by dividing $\frac{57.5}{7.5} \times 0.1 = 0.77$ s. This is the duration of one beat.

To convert to beats per minute divide one minute (60 s) by the time for one beat:

$$\frac{60}{0.77} = 78 \text{ beats per minute.}$$

Measuring the delay between atrial and ventricular systole

This small time delay needs to be long enough to allow atrial systole to complete before the more powerful ventricle contracts. On the ECG this is the PR interval between the start of P and the start of the QRS complex. The length of P–R on the ECG is 15 mm. This is $\frac{9}{7.5} \times 0.1 = 0.12$ s in this example.

Measuring stroke volume

The stroke volume is the volume of blood pumped by one beat. In this example the volume scale shows that the stroke volume is 80 cm^3. This is the difference between the peak volume (ventricle full) and lowest volume (ventricle empty).

Measuring the cardiac output

This is the volume of blood pumped per unit time. In this example the stroke volume is 80 cm^3 and the rate is 102 beats per minute, so the cardiac output is $80 \times 102 = 8.16$ dm^3 per minute (or 8.16 litres per minute).

REMEMBER: The shape of the trace is constant, so if R–R is not visible then measure another suitable interval, e.g. P–P.

SUMMARY QUESTIONS

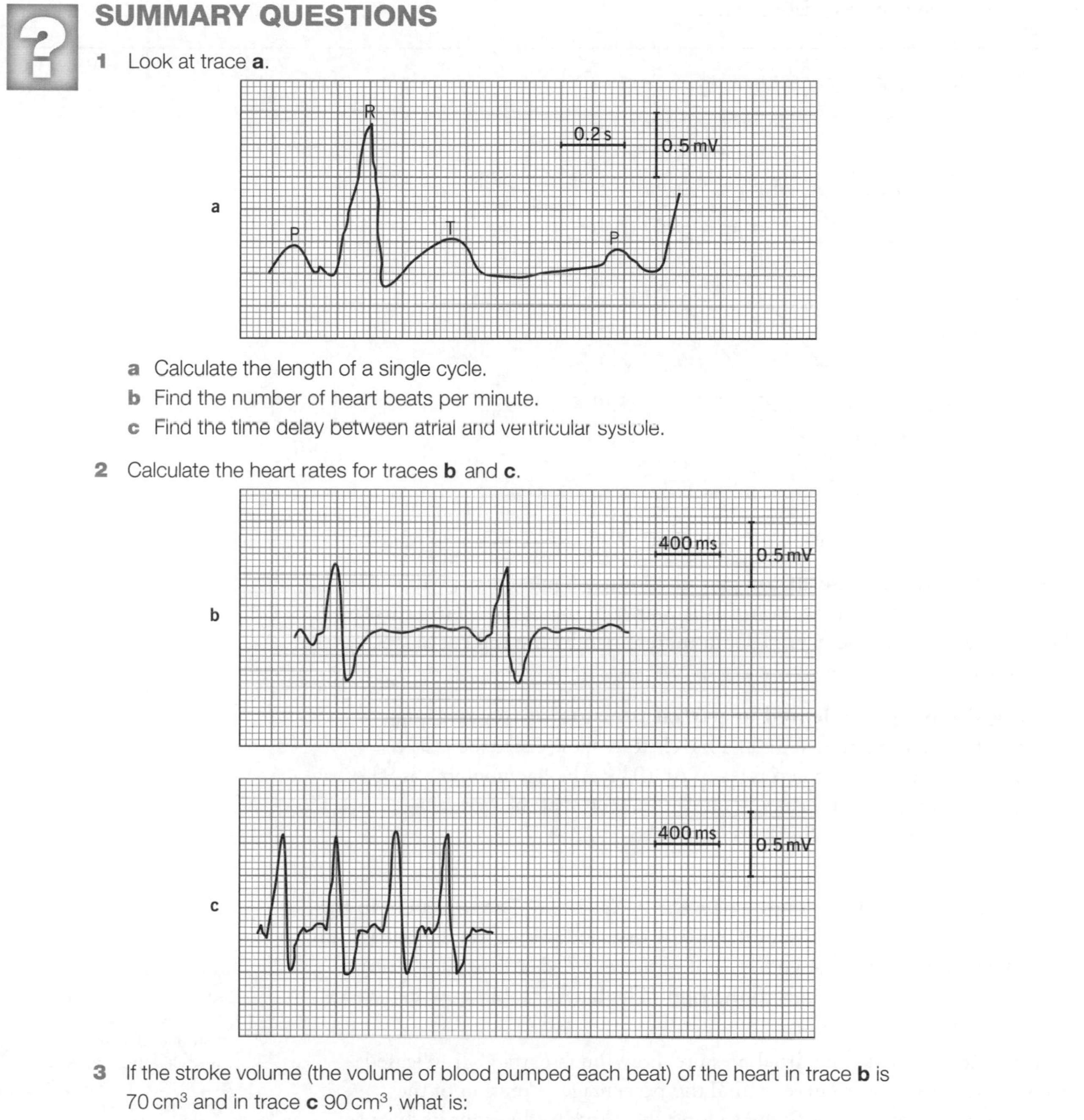

1 Look at trace **a**.

  a Calculate the length of a single cycle.
  b Find the number of heart beats per minute.
  c Find the time delay between atrial and ventricular systole.

2 Calculate the heart rates for traces **b** and **c**.

3 If the stroke volume (the volume of blood pumped each beat) of the heart in trace **b** is 70 cm^3 and in trace **c** 90 cm^3, what is:

  a the cardiac output (volume of blood pumped per minute) for each trace
  b the percentage increase in the cardiac output of the heart between traces **b** and **c**?

4 Look back at the first example on page 66. Use the pressure trace to identify the exact times when the following events occur:

  a opening of the atrio-ventricular valves
  b closing of the semi-lunar valves.

Oxygen dissociation curves

Understanding oxygen dissociation curves

As oxygen partial pressure increases, more oxygen combines with haemoglobin until the haemoglobin is saturated. The relationship between oxygen partial pressure and percentage saturation of haemoglobin can be plotted as an oxygen dissociation curve.

WORKED EXAMPLE

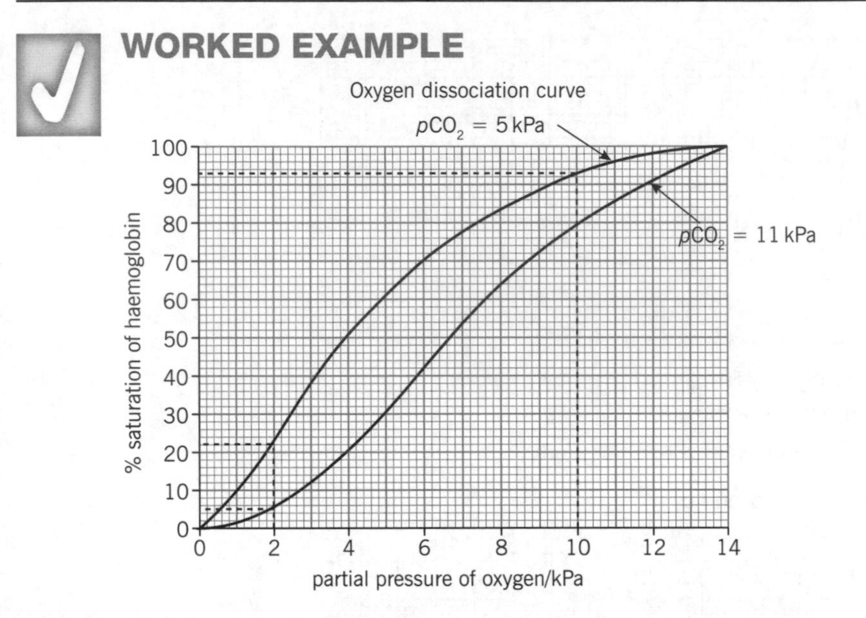

Finding the percentage of oxygen at a named partial pressure
For example, at 10 kPa, draw construction lines to the curve then across to the *y*-axis as shown. The haemoglobin represented by the top curve is 93% saturated at 10 kPa.

Finding the oxygen released as partial pressure of oxygen falls
For example haemoglobin at 10 kPa travels to tissues where the pO_2 is only 2 kPa. What is the oxygen release? At 10 kPa the haemoglobin is 93% saturated. Using the same principles it can be seen that the percentage saturation at 2 kPa is 22%. So the release is $93 - 22 = 71\%$.

The Bohr effect
Haemoglobin loads in the lungs where the pCO_2 is low. The top curve in the diagram illustrates this. In tissues the pCO_2 rises and this causes the behaviour of the haemoglobin to change so that it has a **reduced affinity** for O_2 and so the dissociation curve shifts to the right. The haemoglobin now behaves as shown in the lower curve.

In tissues at 2 kPa oxygen partial pressure, how much extra O_2 is released as result? Use the lower curve to find the percentage saturation of the haemoglobin; here it is 5%. So by shifting the curve to the right in this way $22 - 5 = 17\%$ extra oxygen is released (dissociated) from the haemoglobin.

REMEMBER: Questions can refer to the Bohr shift due to increase in the pCO_2, but reducing the pH or increasing the temperature also shifts the curve to the right.

SUMMARY QUESTION

1 Curves **a**, **b** and **c** represent dissociation curves at 3.0, 5.0 and 8 kPa CO_2 respectively.

a Find the percentage of oxygen released by haemoglobin loading in the lungs at pCO_2 3.0 kPa and pO_2 14 kPa when it reaches tissues at pCO_2 5.0 kPa and pO_2 2.0 kPa.

b How much extra oxygen would be released in the same tissues if the pCO_2 rose to 8 kPa?

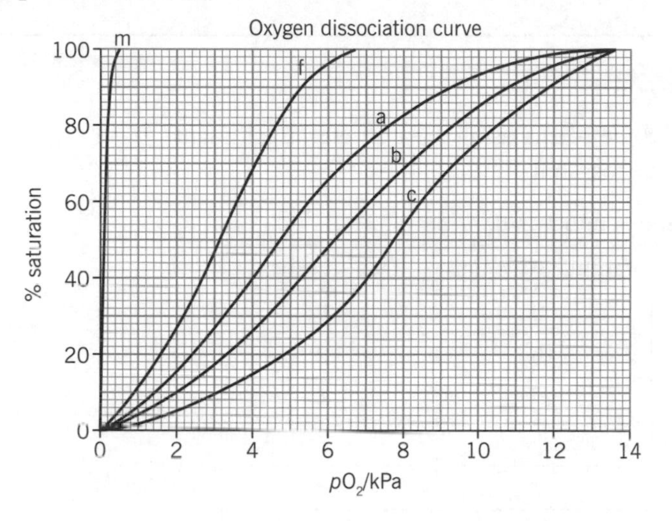

c Curve **f** represents fetal and curve **a** represents maternal haemoglobin. How much extra oxygen can be held by fetal haemoglobin compared with maternal haemoglobin at pO_2 5 kPa?

d If fully saturated blood in curve **a** carries 18 cm^3 per 100 cm^3 of blood, what would be the actual volume of oxygen released by 3 litres of blood when this blood reaches tissues at pCO_2 8 kPa and pO_2 2 kPa?

e Curve **m** represents myoglobin. In normally respiring tissues at pO_2 3 kPa the myoglobin is fully saturated. What is the percentage decrease in pO_2 required to cause full unloading of the myoglobin in these tissues?

Water uptake rate

Measuring water uptake rate using a potometer

Transpiration is the loss of water vapour by diffusion through the stomata in the leaves, where water is also used for reactions such as photosynthesis. Water used in biochemical reactions or lost from the leaf is balanced by uptake from the soil and translocation up the xylem. A potometer can be used to calculate how fast water is being taken up by leafy shoots.

WORKED EXAMPLE

Mike collected data from three plants using a potometer. Each plant originated from a different ecosystem.

Time/s	Meniscus position/mm		
	Tropical plant	Temperate plant	Desert plant
0	0	0	0
30	18	10	2
60	44	22	4
90	66	32	4
120	86	46	6
150	106	52	8
180	134	66	9
210	152	74	11
240	176	86	12

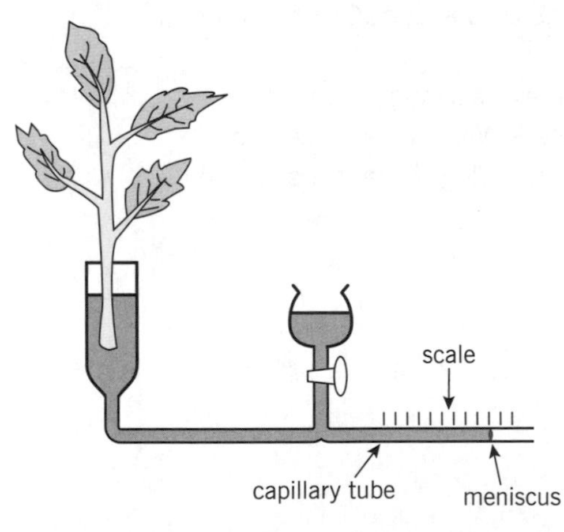

To find the rate of water uptake it is necessary to calculate the volume of water taken into the shoot and the time taken. This can be done directly from the table or by using a graph of the data.

Using the data in the table, the tropical plant results in a meniscus movement of 176 mm in 240 s. The bore of the capillary tube is cylindrical, so the distance moved by the meniscus is equivalent to the length, h, of a cylinder. To find the volume of water absorbed, apply the equation for the volume of a cylinder, which is $\pi r^2 h$ where h = the height or length of the cylinder and r is the radius of the cylinder.

REMEMBER: Always begin by plotting the graph because the rate may vary. When this is the case it is necessary to select the range of the graph from which to calculate, e.g. where the line is steepest to give the maximum rate.

The potometer is not a direct measure of transpiration!

Mike used a capillary tube with a bore 1 mm in diameter (i.e. radius 0.5 mm), so the volume of water taken up by the tropical plant was $\pi r^2 h$ or $3.14 \times 0.5^2 \times 176 = 138.2\,\text{mm}^3$.

Convert this to a rate by dividing by the time taken, $\frac{138.2}{240} = 0.58\,\text{mm}^3\,\text{s}^{-1}$.

SUMMARY QUESTIONS

1 Plot a suitable graph to include the data from all three plants in Mike's experiment.

2 Find the rate of water uptake by the plants from the temperate and desert ecosystems using Mike's data in the table.

3 In another potometer with tube internal diameter 3 mm the meniscus moved a total of 16 mm in 120 s. What is:
 a the total volume of water taken into the plant
 b the rate of uptake of water?

4 Using a potometer with tube internal diameter 1 mm the rate of water uptake during a period of 4 minutes was found to be $1.2\,\text{mm}^3\,\text{s}^{-1}$. What was the total distance moved by the meniscus in this experiment?

Oxygen release from photosynthesis

Measuring oxygen release from photosynthesis

If an apparatus is used to collect the oxygen bubbles produced by a sample of an aquatic plant then the volume of gas can be measured, which is much more precise than simply counting the bubbles. The principle of calculating volume is the same as for the potometer or respirometer because a tube of known internal diameter is used to calculate the volume of the gas.

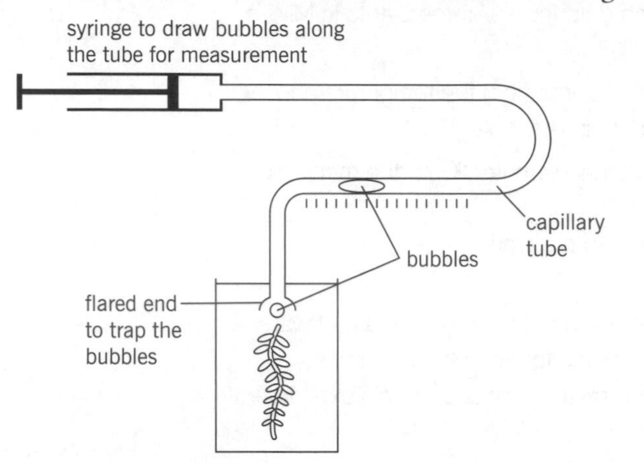

WORKED EXAMPLE

Lamp distance d/cm	Light intensity $\left(\frac{1000}{d^2}\right)$	Bubble lengths collected in 60 s/cm				Bubble volumes collected in 60 s/mm³			
		1	2	3	mean	1	2	3	mean
15	4.44	7.9	8.7	8.6	8.4	62.0	68.3	67.5	65.9
20		7.9	7.7	7.9					
25		6.6	6.8	6.4					
30		5.1	5.2	4.8					
35		4.2	4.1	4.2					
40		2.7	2.8	2.7					

A photosynthometer was used to collect the oxygen from a 40 g sample of *Elodea* pondweed while varying the illuminating lamp distance. The internal diameter of the photosynthometer tube was 1 mm.

Calculating the light intensity

Light intensity is inversely proportional to the square of the distance, d. This is called the inverse square law. Mathematically this is represented as intensity $\propto \frac{1}{d^2}$. To convert distances of the lamp into intensities the simplest method is to say intensity $= \frac{1}{d^2}$, but this gives very small numbers, so it is easier to use $\frac{1000}{d^2}$ to calculate the values. At 15 cm the intensity is

$$\frac{1000}{15^2} = 4.44$$

No units are applied in this case as the intensities calculated are relative measurements only.

Calculating the bubble volumes

Each bubble is trapped in a cylindrical tube, so its volume is found by using the formula $\pi r^2 h$. In this formula r is the radius of the tube and h is the length of the bubble.

Remember to convert the units at the start. So 7.9 cm = 79 mm and the oxygen volume is $3.14 \times 0.5^2 \times 79 = 62.0\ \text{mm}^3$. In this example each bubble is calculated individually to give an opportunity to practise, but the mean could just as easily be used.

Calculating the rate of photosynthesis

Now the volume of gas and the time is known, find the rate by dividing volume by time. At 15 cm lamp distance the rate is $\frac{65.9}{60} = 1.10\ \text{mm}^3\,\text{s}^{-1}$.

Finding the optimum light intensity

This step involves plotting the rates against light intensity. Using the graph, the lowest light intensity that gives the maximum rate can be read as shown in the sketch. This is the optimum light intensity.

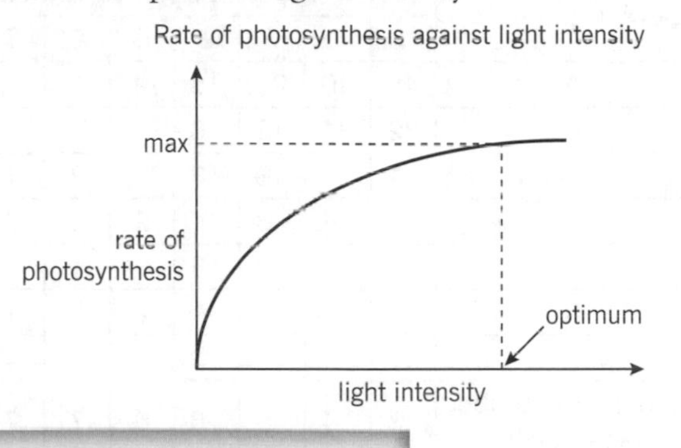

REMEMBER: Whenever tubes such as a capillary tube are used to measure, the formula $\pi r^2 h$ will always allow you to convert bubble length or meniscus movement into volumes. Be careful not to mix units when calculating.

SUMMARY QUESTIONS

1. Calculate the values and complete the table to show:
   - **a** the light intensities
   - **b** the mean bubble lengths
   - **c** the bubble volumes.
2. Calculate the rates of photosynthesis for each light intensity as $\text{mm}^3\,\text{s}^{-1}$.
3. Plot a graph to show rate against light intensity and suggest why it is not possible to find the optimum light intensity. What experimental changes would allow the optimum to be identified?
4. A seventh lamp distance was added to the experiment, which was repeated exactly as before collecting the gas for 60 s. The calculated rate turned out to be $0.28\ \text{mm}^3\,\text{s}^{-1}$. Find:
   - **a** the mean bubble volume that was collected each 60 s
   - **b** the mean length of the bubble collected in 60 s.

Quadrat data

Interpretation of quadrat data

Quadrats are frames of fixed size that are used to gather samples. These quadrat samples are considered to be subsets of a total sampling area. If they are used skilfully they can measure a variety of ecological parameters. The size of a quadrat may be varied to suit the organisms being investigated. A 10 × 10 cm quadrat could be useful when sampling lichen patches on a wall, while a 0.05 × 0.05 mm square on a haemocytometer is suitable for sampling cells on a microscope slide. Most often you will be using a 50 × 50 cm quadrat and sampling plants or animals, e.g. on lawns or rocky shores.

WORKED EXAMPLE

Checking that all organisms in the sample area have been seen

Position the first quadrat using random number coordinates. Record the species encountered in a list.

Continue sampling, adding new species encountered to the list as you go.

Once no new species are encountered in successive quadrats, you will have seen all the species once (possibly not the extremely rare ones!). If a careful preliminary walking survey of your sample site has allowed you to spot the various species it will be easier to be certain that the sampling procedure has measured them all.

Species	% cover in each 50 × 50 cm quadrat									
	1	2	3	4	5	6	7	8	9	10
A	45	40	21	15	9	2				
B	32	31	12	8	2					
C	1		18	8		3	1			
D		4	3	1	2	2	7	1		
E			2	6			1			
F				1	2	1	1	2	5	6
G						3	4	16	21	19
Cumulative number of species seen	3	4	5	6	6	7	7	7	7	7

The number of quadrats required is called the **critical number**. This is easily shown if a sketch graph is made of the cumulative number of species encountered against the number of quadrats sampled.

In the table, seven species have been sampled. After six quadrats no new species are seen, so the critical number is six for this site.

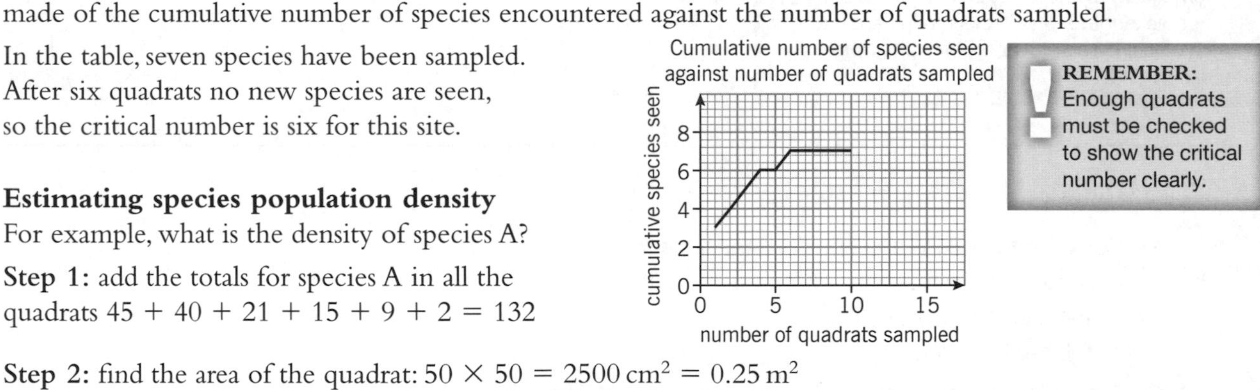

REMEMBER: Enough quadrats must be checked to show the critical number clearly.

Estimating species population density

For example, what is the density of species A?

Step 1: add the totals for species A in all the quadrats 45 + 40 + 21 + 15 + 9 + 2 = 132

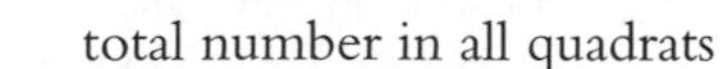

Step 2: find the area of the quadrat: $50 \times 50 = 2500\text{ cm}^2 = 0.25\text{ m}^2$

Step 3: calculate density using $\dfrac{\text{total number in all quadrats}}{\text{number of quadrats sampled} \times \text{area of quadrat}} = \dfrac{132}{10 \times 0.25} = 52.8$ per m^2.

In this case you expect an average of 53 per m^2 for species A.

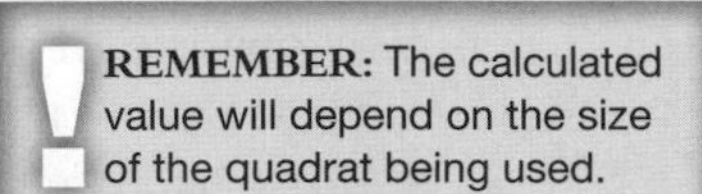

REMEMBER: The calculated value will depend on the size of the quadrat being used.

Finding species frequency

This is the probability of a species occurring in any quadrat.

For example what is the frequency of species D?

Use the calculation

$$\frac{\text{number of quadrats in which the species is seen}}{\text{total number of quadrats}} \times 100 = \frac{7}{10} = 70\%$$

Finding the total population of a species in the sample area

For example, how many worms are there in an area of grassland 400 m long by 176 m wide?

Step 1: take samples from 0.5 × 0.5 m quadrats placed at random within the area studied.

Quadrat	1	2	3	4	5	6	7	8	9
Number of worms	2	16	4	21	14	1	14	18	4
Cumulative total	2	18	22	43	57	58	72	90	94
Mean number of worms	2	$\frac{18}{2} = 9$	$\frac{22}{3} = 7$	$\frac{43}{4} = 11$	$\frac{57}{5} = 11$	$\frac{58}{6} = 10$	$\frac{72}{7} = 10$	$\frac{90}{8} = 11$	$\frac{94}{9} = 10$

Step 2: calculate the cumulative mean number of worms found. To start with this number will fluctuate but it will gradually stabilise. This takes into account the different dispersal of the worms; some areas have dense and some sparse populations. Once the mean stabilises (this can be shown on a sketch graph) you can be confident that an adequate sample has been gained. Here the mean appears to be 10, but a few more samples could confirm this.

Step 3: find the area of the sample site: $400 \times 176 = 70\,400\ \text{m}^2$

Step 4: calculate the area of a quadrat: $0.5 \times 0.5 = 0.25\ \text{m}^2$

Step 5: calculate the total worms in the whole area:

$$\frac{\text{total area of site}}{\text{area of one quadrat}} \times \text{mean number per quadrat} = \frac{70\,400}{0.25} \times 10$$

$= 2\,816\,000$ worms in the whole field.

REMEMBER: Be careful in situations where a species occurs in clumps (underdispersed).

SUMMARY QUESTIONS

The table shows data collected from an area of rocky shore measuring 48 m wide by 12 m deep.

	Number of animals found in each 0.25 m² quadrat								
Species	1	2	3	4	5	6	7	8	9
Beadlet anemone	0	0	1	0	2	3	1	0	3
Flat periwinkle	0	0	1	6	2	18	2	4	4
Grey top shell	15	0	4	11	9	8	5	0	1
Purple top shell	12	7	9	3	0	3	12	34	16
Dog whelk	2	6	4	0	6	3	0	0	2
Chiton	0	0	0	1	0	0	0	1	0

1. What is the frequency of:
   - **a** the flat periwinkle
   - **b** the chiton?
2. What is the critical number of quadrats?
3. What is the species density of:
   - **a** the dog whelks
   - **b** the purple top shells?
4. In the area studied, what is the population of:
   - **a** the grey top shells
   - **b** the beadlet anemones?
5. What is the probability that a chiton will be found in the 10th quadrat?

Species diversity

Calculating species diversity

Species diversity should not be confused with species richness, which is the total number of species present in a community. Diversity is a calculated measure of the balance between the different species present. If diversity is high it suggests that the community is evenly balanced and likely to be stable, e.g. ancient woodland. Low diversity can be indicative of communities which have over representation by some species and may be liable to change as the years pass. For example, the early stages of succession such as pioneer communities. Low values may also indicate dominant populations in specialised habitats, for example barnacles on rocks in the upper shore.

WORKED EXAMPLE

Data required from which to calculate the diversity includes the numerical abundance for each individual species (n) and the total numerical abundance of all species together (N). For plants, numerical abundance can be substituted for percentage cover data. The example below compares two sites.

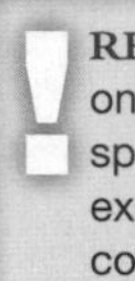

REMEMBER: There is more than one arrangement of the formula for species diversity, depending on your exam board. Examples of the two common versions are shown below.

	Calculation for site 1			Calculation for site 2		
Plant species	**Number at site 1 (n)**	$\frac{n}{N}$	$\left(\frac{n}{N}\right)^2$	**Number at site 2 (n)**	$\frac{n}{N}$	$\left(\frac{n}{N}\right)^2$
A	84	$\frac{84}{100} = 0.84$	0.71	22	$\frac{22}{100} = 0.22$	0.05
B	2	$\frac{2}{100} = 0.02$	0.00	30	$\frac{30}{100} = 0.30$	0.09
C	6	$\frac{6}{100} = 0.06$	0.00	25	$\frac{25}{100} = 0.25$	0.06
D	8	$\frac{8}{100} = 0.08$	0.00	23	$\frac{23}{100} = 0.23$	0.05
Totals	$N = 100$		0.71	$N = 100$		0.25
Species richness	4			4		

Diversity is measured by calculating the value of Simpson's Index of Diversity, D. This value does not have units but is a fraction of one. If all species had exactly the same populations then the balance between them is equal and D will be 1. The more uneven the numbers of species the smaller D becomes.

Method 1 (used by e.g. OCR)

The formula required is $D = 1 - \sum\left(\frac{n}{N}\right)^2$

Placing the data in a table like the one above greatly facilitates the calculation.

For site one $D = 1 - 0.71 = 0.29$

For site two $D = 1 - 0.25 = 0.75$

Note that for small n numbers the values of $\left(\frac{n}{N}\right)^2$ can often be rounded to zero.

Method 2 (used by e.g. AQA)

The formula required is $d = \dfrac{N(N-1)}{\sum n(n-1)}$

Plant species	Number at site 1 (*n*)	*n*(*n* − 1)	Number at site 2 (*n*)	*n*(*n* − 1)
A	84	84 × 83 = 6972	22	22 × 21 = 462
B	2	2 × 1 = 2	30	30 × 29 = 870
C	6	6 × 5 = 30	25	25 × 24 = 600
D	8	8 × 7 = 56	23	23 × 22 = 506
Totals	*N* = 100	Σ*n*(*n* − 1) = 7060	*N* = 100	Σ*n*(*n* − 1) = 2438
N(*N* − 1)	100 × 99 = 9900		100 × 99 = 9900	

Site 1 $d = \dfrac{9900}{7060} = 1.4$

Site 2 $d = \dfrac{9900}{2438} = 4.1$

Again, when using this method, the higher the value of *d*, the greater the diversity.

The same conclusions may be drawn using either method. Although species richness and total number of plants present suggest that the sites are the same, the diversity clearly reveals the difference in balance. Site two has a higher diversity and this indicates more uniform balance between the species present. Site one, with a lower diversity, could be a cornfield dominated by wheat, or a sand dune dominated by marram grass. Site two could be an established woodland or a conservation meadow.

REMEMBER: Collect the original data using random sampling.

This calculation will not work if biased data is collected!

SUMMARY QUESTION

1 Three examples of data are shown in the table below.

a Calculate the species diversity for each site.
Use the formula matching that required by your syllabus!

b What is the species richness for each site?

Meadow		Rocky shore		Wheat field	
Species	% cover	Species	Number found	Species	% cover
Buttercup	7	Limpet	76	Wheat	99
Daisy	16	Purple top shell	14	Wild oat	3
Meadow grass	38	Toothed top shell	49	Clover	13
Yarrow	6	Edible periwinkle	35	Nettle	9
Self heal	5	Chiton	3	Dock	2
White clover	26	Beadlet anemone	13		
Speedwell	3	Grey sea slug	2		
Plantain	8	Dog whelk	14		
Dandelion	11	Barnacle	167		
Rye grass	23				

Species richness and species evenness

Assessing species richness and species evenness

Species richness is the total number of species represented within a sample area. However, it gives no indication of how many of each species are present. Species evenness is the relative abundance of the different species that are present. It gives valuable information about the numerical balance between the species. In studies of ecosystems, species evenness may be monitored through time to track the establishment of new species or the loss of species. It can be used to assess interspecific competition.

Calculating species evenness is complicated; use it to stretch and challenge yourself!

WORKED EXAMPLE

To begin assessing species richness and evenness, data need to be collected using random quadrat sampling within the study area. The number of individuals of each species present is n. The total number of individuals of all species present is N. It is a lot simpler to collect the data directly into a table, for example:

Species name	Number present (n)
Bluebell	38
Celandine	46
Wood anemone	21
Total for all species (N)	105

The species richness, S, is easily determined by counting the number of species listed in the first column, here it is 3.

Species evenness E_H is found using this formula $E_H = \dfrac{H}{\ln S}$

H is Shannon's diversity index and is found using this formula:

$$H = -\sum Pi(\ln Pi)$$

Pi is the proportion of each species and is found by dividing $\frac{n}{N}$ for each species.

$\ln Pi$ is the natural log of Pi.

Note the minus sign, which means multiply $\sum Pi(\ln Pi)$ by -1 at the end (this effectively changes the sign).

Using a table is helpful in performing the calculation of H.

Species name	Number present (n)	Proportion of each species (Pi)	$\ln Pi$ for each species	$Pi(\ln Pi)$ for each species
Bluebell	38	$38 \div 105 = 0.36$	-1.02	$0.36 \times -1.02 = -0.37$
Celandine	46	$46 \div 105 = 0.44$	-0.82	$0.44 \times -0.82 = -0.36$
Wood anemone	21	$21 \div 105 = 0.20$	-1.61	$0.20 \times -1.61 = -0.32$
Total for all species (N)	105			$\sum Pi(\ln Pi) = -0.37 + -0.36 + -0.32 = -1.05$ so $H = -1.05 \times -1 = 1.05$

The species richness is 3 and the natural log $\ln S$ of this is 1.10.

So the species evenness

$$E_H \text{ is } \frac{H}{\ln S} = 1.05 \div 1.10 = 0.96$$

Species evenness is always on a scale between zero (total lack of evenness) and one (completely even). In this example the figure of 0.96 suggests that the species evenness is high (close to 1) and therefore all species are equally represented.

REMEMBER:

Use a suitable sample size

If your sample is too small then large errors will appear in your calculations. For example, rare species may not be seen at all.

It is also possible to calculate by using populations estimated by averaging sampled quadrats, or even by using percentage cover data such as might be gathered for plants.

SUMMARY QUESTION

1 Calculate the values of species richness (S) and species evenness (E_H) for the samples below, which were collected for the same area in different years. The data show fluctuations in populations of three bird species in southeast Sweden. (Source Hjort, C. and Lindholm, C.G. 1977. Oikos 30: 387–392.)

Species	Sample years			
	1946	1956	1966	1976
Wren	5	10	90	170
Whitethroat	118	140	150	100
Yellowhammer	8	78	68	130

Transects and kite diagrams

Transects and kite diagrams

When quadrat samples are taken along an oriented transect line, the data can be presented using a kite diagram. Kite diagrams can demonstrate the variations in species richness, relative abundance and abiotic factors across the sample area.

WORKED EXAMPLE

Alistair and Caro collected data from the edge of a piece of woodland.

Species name	Distance along the transect/m					
	0	5	10	15	20	25
	Percentage cover by each species					
Silver weed	10	2	0	27	0	0
Grass	80	87	92	71	14	0
Dandelion	15	1	2	0	0	0
Ground ivy	0	0	0	6	67	78
	Environmental factors					
Light intensity/arbitrary units	82	85	78	67	45	20
Soil pH	6.2	6.2	6.3	5.8	5.0	4.5
Other factors	path edge	path	path	woodland edge	woodland edge	woodland

To plot data on a kite diagram the abundance at each point should be plotted symmetrically on either side of a guide line, otherwise the 'kite' shapes will not emerge neatly. Begin at the bottom of the graph, draw a guide line that allows enough space either side to accommodate 50% of the range to be plotted above and below the line. Mark the limits of the plot, e.g. for grass at site one (0 m along the transect) 40% above and 40% below the guide line.

The abiotic data may be plotted above or below the data for the organisms observed.

Species richness
This is the total number of species seen.

Species diversity and evenness
These can be calculated at each station using the methods on pages 76 and 78 and plotted as line graphs to show variation across the transect.

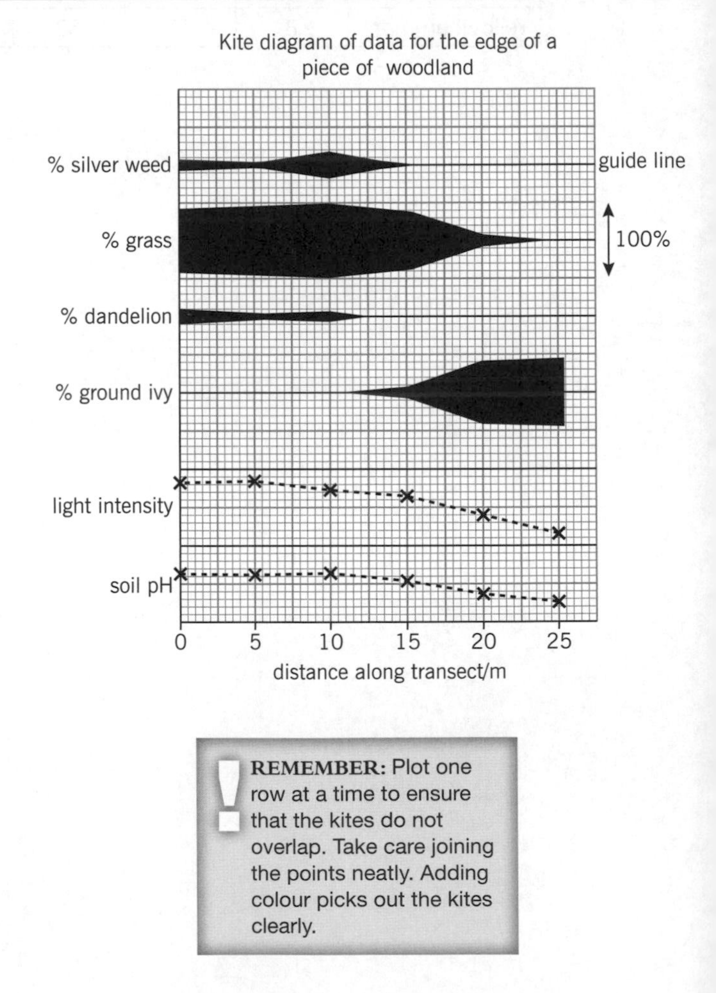

REMEMBER: Plot one row at a time to ensure that the kites do not overlap. Take care joining the points neatly. Adding colour picks out the kites clearly.

STRETCH YOURSELF

Trying an alternative abundance scale

What about plotting large ranges of numbers for many organisms?

When a large number of kites need to be plotted it may not be feasible to fit the data on the available *y*-axis. When this is the case, it is possible to convert percentage cover data to a simpler format. One way of doing this is to use the Domin scale, which also helps with the consistency of observations because it uses ranges to accommodate subjective percentage cover judgements. All abundances are scaled between 1 and 10, so the number of plots that can be fitted on the *y*-axis is increased.

Cover (%)	Domin value
91–100	10
76–90	9
51–75	8
34–50	7
26–33	6
11–25	5
4–10	4
< 4 with many individuals	3
< 4 with several individuals	2
< 4 with few individuals	1

SUMMARY QUESTION

1 Construct a kite diagram using the following data. Calculate species diversity and richness at each station (distance) then show the trends with a line on the kite diagram. There is space in the bottom rows of the table for your calculated values.

Distance inland/m	10	20	30	40	50	60	70	80	90	100	110	120	130	140	150	160	170	180
Species	% cover data																	
Bare sand	70	50	15	10	55	55	22	20		50	42	8		40	12	24	20	
Marram grass	40	85	90	70	75	70	44	30	10									
Sand sedge				10	15		8		10		2							
Sea couch grass		5		3	3		6			2		1						
Dandelion							4	10	6	13			10		10		4	
Lyme grass	1	4		4	8	3							70		20	15	32	20
Sea spurge		15		4			1											
Sea holly					6		2											
Cat's ear		7		4	10			3		5			50					
Ragwort		1	6	7											15			
Hawkweed				2		3				10						5		
Sea rocket				15		28	10											
Lesser hawkbit				4	13	1		10	10		5	20	10	5	16	2	10	
Moss													4	8	1	1	1	5
Red fescue grass						20	20		40	30		20	10	40	75	5	18	60
Sea bindweed						23	19											
Ribwort plantain						4			8						16	40	27	
Rest harrow										44		40	40		52	33	32	40
Portland spurge										4		6	10	1				
Common centaury										45				3		8		5
Sheep's fescue										20				20		40		30
Eyebright											10		25		14	12		
Clover											100	8		20	30			
Bird's foot trefoil												40	80		60	35	5	90
Pyramidal orchid															4			8
Species richness																		
Species diversity																		

Mark–release–recapture

Counting species by using mark–release–recapture

Species that are mobile can be counted using a technique called mark–release–recapture, sometimes also called capture–mark–recapture or simply mark–recapture. The method involves making an estimate of the population size by calculating the Lincoln Index. Animals are captured, counted, marked then released randomly back into the study site. After a time they mix evenly with the unmarked portion of the total population. A second sample is taken and the number of marked and unmarked individuals is recorded. The calculation is a ratio assuming equality between the total marked and the total unmarked individuals in the population.

WORKED EXAMPLE

Bridget set pitfall traps randomly in a deciduous woodland. After 24 hours she removed and counted all the violet ground beetles in the traps and recorded the total (S_1). She marked every animal indelibly and unobtrusively then released them back where they had been collected. Two days later she reset the traps in the same place and collected a second sample. She counted the total number of beetles in this new sample (S_2) and also recorded how many of the recaptured animals were marked (R).

The population, N could then be calculated using the Lincoln Index equation

$$N = \frac{S_1 \times S_2}{R}$$

The assumption being made here is that $\frac{S_1}{N} = \frac{R}{S_2}$, i.e. that the ratio of marked to unmarked beetles in sample 2 is the same as the ratio of the original sample to the whole population. This can only be an estimate as it assumes complete proportional mixing of the marked with the unmarked part of the population and also that no change has occurred in the population between sample sessions. It is also assumes that no marks have worn off!

In Bridget's first sample she caught and marked 56 beetles ($S_1 = 56$). She captured 67 beetles in the second sample ($S_2 = 67$). Of these 67 she found that 12 were marked ones from the original sample ($R = 12$).

The calculation was therefore $\frac{56 \times 67}{12} = \frac{3752}{12} = 313$

So the estimated population was 313 beetles.

REMEMBER:
This method assumes that the population remains unchanged during the process, that marking favours neither survival nor death nor predation and that immigration and births equal emigration and deaths.

SUMMARY QUESTION

1 The table shows seven examples of data collected using the mark–release–recapture technique. Calculate the missing values in each case.

		S_1	S_2	R	N
a	Woodlice under a rotting tree trunk	23	28	6	
b	Soldier beetles in a length of hedgerow	127	44	2	
c	Banded snails in a vegetable plot	96		21	347
d	Ground beetles in a garden		88	5	1707
e	Sand hoppers in seaweed on the strandline	234	186		14 508
f	Short-tailed voles in an area of grassland	145		17	1049
g	Smooth newts in a pond		8	1	96
h	Purple topshells on an area of rocky shore	102	98		1428

Population growth curves 1

Interpreting population growth curves 1

Once a complete growth curve is plotted, it quickly becomes apparent that there is a maximum population, after which numbers remain constant. The exponential part of the curve will also have a gradient related to the speed of increase of the population, faster increase giving steeper slopes. The relationship between the maximum population and the rate of exponential growth may be shown mathematically.

WORKED EXAMPLE

The maximum stable population is called the carrying capacity, represented by K.

A typical population growth curve

The slope of the exponential phase depends on a factor called r, the **intrinsic rate of natural increase** or **biotic potential**. Quickly multiplying species have high values of r and are called r-selected. They can colonise new areas quickly.

The rate of increase in the population during the exponential phase depends only on r because there are no limiting factors. So the rate of increase, R, can be calculated using:

$R = \frac{dN}{dt} = rN$, where N is the number in the population.

For example, if the population rises from 35 to 65 in 1 month, dN (the change in numbers) must be $65 - 35 = 30$, so $R = \frac{30}{1} = 30$ individuals per month.

As the population approaches the carrying capacity, the ability of the population to increase becomes more and more limited. The equation of population growth is modified to take into account the value of K, becoming:

$$\frac{dN}{dt} = rN\left(1 - \frac{N}{K}\right)$$

or

$$R = rN\left(1 - \frac{N}{K}\right)$$

For example, if $K = 670$ and $r = 12$ then when the population is 167,

$R = 12 \times 167\left(1 - \frac{167}{670}\right) = 1504.5$ individuals per unit time; a rapid growth rate.

But when the population reaches 590 this changes so that:

$R = 12 \times 590\left(1 - \frac{590}{670}\right) = 845$ individuals per unit time; a slower growth rate.

When the population reaches the carrying capacity:

$R = 12 \times 670\left(1 - \frac{670}{670}\right) = 0$ individuals per unit time;

the population is at the maximum stable size.

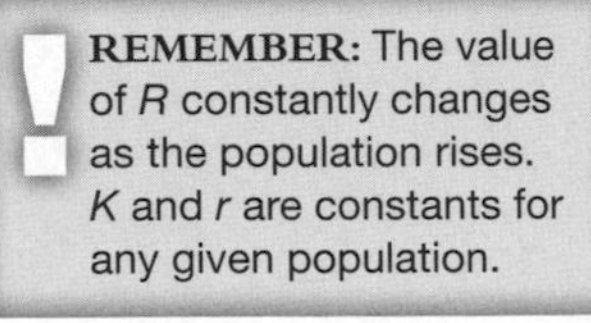

REMEMBER: The value of R constantly changes as the population rises. K and r are constants for any given population.

SUMMARY QUESTIONS

Intrinsic rate of natural increase (*r*)	Number of individuals in the population (*N*)	Carrying capacity (*K*)	Rate of population increase (*R*)
7	2	92	13.7
7	4	92	
7		92	51.1
7	12	92	73.0
7	16	92	
7	20	92	109.6
7	24	92	124.2
7	28	92	136.3
7		92	146.1
7	36	92	153.4
7	40	92	
7	44	92	160.7
7		92	160.7
7	52	92	158.3
7	56	92	153.4
7	60	92	146.1
7	64	92	136.3
7		92	124.2
7	72	92	109.6
7	76	92	92.5
7	80	92	
7	84	92	51.1
7	88	92	26.8
7	92	92	0.0

The table illustrates some data on the growth of a population.

1. Copy the table and fill in the missing values of *N*.
2. Calculate the missing values of *R*.
3. Plot a graph of *R* against *N*.
4. Explain the shape of the graph.
5. If the food supply increased so that *K* rose to 178, what would be the rate of population growth when the population reached 102 individuals?

Population growth curves 2

Interpreting population growth curves 2

If a population is monitored over time, the number of organisms can be plotted to produce a typical sigmoid growth curve. The growth of populations during the exponential phase can be predicted mathematically and a number of quantities can be calculated by using the curve.

WORKED EXAMPLE

Time/min	Division number	Population/no. cells	Log_{10} no. cells
0	0	1	0
20	1	2	0.301
40	2	4	0.602
60	3	8	0.903
80	4	16	1.204
100	5	32	1.505
120	6	64	1.806

The table above shows data taken from a growing culture of bacteria. The cells reproduce by binary fission. Since they divide more or less simultaneously, each division in column 2 refers to the whole population. Thus the numbers in the population column double with each division.

Finding the generation time

During the growth of bacteria the generation time is the time it takes for the population to double. This is simply read off the table; in the example above it is 20 minutes.

How to predict the growth of the culture

Once the generation time is known it is possible to calculate the population at any given time using the equation $N_t = N_0 \times 2^n$

N_t = the number of cells at time t.

N_0 = the initial number of cells.

n = the number of generations in time t.

For example at 100 minutes $n = 100 \div 20 = 5$

As the population began with one cell $N_0 = 1$

So the number of cells after 100 minutes $N_t = 1 \times 2^5 = 32$

Finding the growth rate constant

The growth rate constant, k, is a numerical value describing the growth of an organism in generations per unit time $k = \frac{n}{t}$

For example for the data above $k = \frac{6}{2} = 3$ generations per hour.

REMEMBER: This relationship applies only during the exponential growth phase.

SUMMARY QUESTIONS

1. Find N_t for a bacterial culture after 2 hours of growth where the initial population was 20 cells and the generation time is 12 minutes.

2. In a culture of bacteria the initial population was 1500 cells. After 8 generations the population was 3^{11} cells.
   - **a** The time taken for the population to reach 3^{11} cells was 4 hours. What was the generation time?
   - **b** What is the growth rate constant for this organism?

3. An initial inoculum of 10^6 bacteria was incubated at optimum temperature. The growth rate constant for this organism is 3 generations per hour. If food supply was not a limiting factor, predict the population of bacteria after 24 hours.

Energy flow between trophic levels

Illustrating energy flow between trophic levels

The solar energy incident on living systems is not all fixed as carbohydrate. That which is may be passed from the plant producers up the food chain when the plants are eaten. Since energy is lost or utilised at each trophic level, the amount passed between successive levels diminishes until no further levels can be supported. The flow of energy through the food chains and webs can be illustrated using a Sankey diagram.

WORKED EXAMPLE

Consider an example where the incident solar energy is fixed as plant biomass in net primary productivity.

	Energy %
Arriving at the leaf surface	100
Reflected off the leaf	10
Passing through the leaf	10
Absorbed by chloroplasts	80
Not utilised in photosynthesis (e.g. wrong wavelength/heat loss)	66
Fixed as glucose (gross primary productivity)	14
Used in plant respiration	12.5
Chemical energy in plant structure	1.5

To represent this as a Sankey diagram you need to draw a flow chart in which the energy transfers are represented by proportional arrow widths.

Begin by deciding on a suitable scale. For example, if 100% is represented by an arrow 10 cm wide then 1.5% would require an arrow 1.5 mm wide. This is a reasonable scale for a normal piece of graph paper.

Step 1: draw a 10 cm wide block arrow starting at the top of the paper to represent the incident light.

Step 2: draw a 1 cm arrow diverging from the main arrow to represent the 10% reflected.

Step 3: continue the main arrow but reduced to 9 cm wide.

Step 4: draw a 1 cm arrow diverging from the main arrow to represent the 10% passing through the leaf.

Step 5: continue the main arrow but now reduced to 8 cm wide.

Continue in this manner until all the energy transfers have been represented.

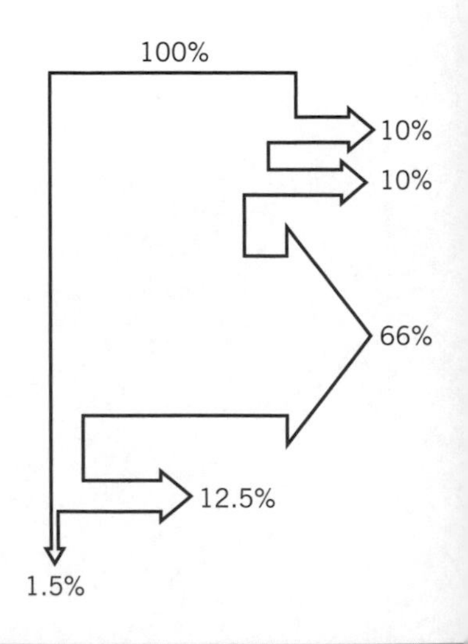

Calculating efficiency of transfer of energy

The efficiency of transfer is $\frac{\text{energy transferred}}{\text{energy intake}} \times 100\%$

In this example it is 1.5%.

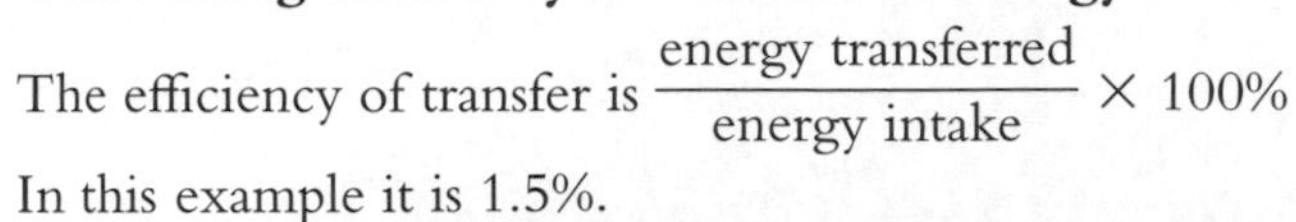

REMEMBER:
In questions the energy flow is sometimes represented by simple lines rather than proportional width arrows. Use the figures not the arrow widths to measure energy flow.

SUMMARY QUESTIONS

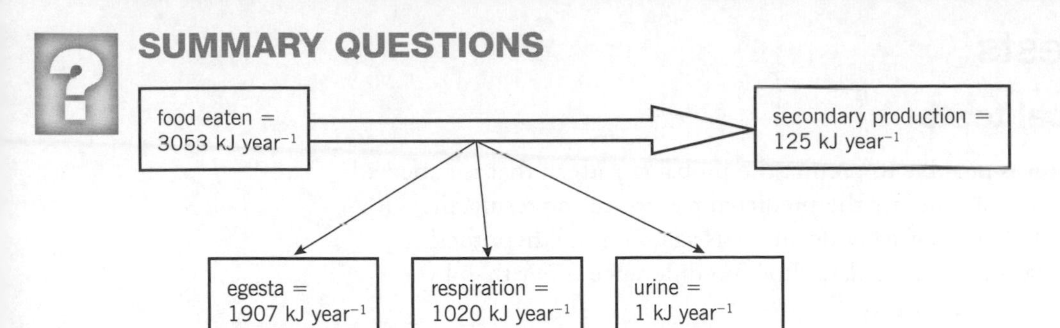

1 The energy flow diagram above represents the energy flow in a bullock.
- **a** What is the efficiency of energy transfer from food to bullock?
- **b** What is the egesta expressed as a percentage of energy intake?
- **c** Suggest ways of increasing the efficiency of energy transfer to bullock biomass.

2 In an ecosystem, the radiant solar energy was 7.1×10^6 kJ m^{-2} year^{-1}. Net primary productivity was 87 403 kJ m^{-2} year^{-1}, energy in tertiary consumers was 88 kJ m^{-2} year^{-1} and 0.05% of the incident energy ended up in primary consumer biomass. The energy in the secondary consumers was 1609 kJ m^{-2} year^{-1}. Calculate the efficiency of energy transfer between:
- **a** the incident radiation and the plant biomass
- **b** primary and secondary consumers
- **c** primary producers and primary consumers
- **d** primary producers and tertiary consumers.

Statistical tests

Using statistical tests

When we collect data it is possible to predict the probable pattern that is expected. We use statistical tests to test whether the predicted pattern in the results fits with the data actually collected. We normally do the tests against a null hypothesis, which is a statement that suggests the data show no difference other than by chance alone.

WHICH STATISTICAL TEST SHOULD BE USED?

Although there is a plethora of tests available, biologists most commonly make use of a specific list of tests because these fit with the type of data that they tend to collect.

Example 1

Resting heart rate in two groups of 30 people before and after a period of training.

Null hypothesis = the heart rates will show no significant difference.

1 The data are ordinal as heart rate is on a measured scale (go to box 2).

2 We are comparing sets of measurements (go to box 3).

3 We are testing the similarity of the means (go to box 5).

4 There are two groups (go to box 7).

5 Heart rate will show a normal distribution (go to box 11).

6 The same individuals are tested before and after (go to box 14).

7 Use the paired Student *t*-test.

Example 2

The longer a person's arms, the faster they will swim.

Null hypothesis = long and short armed people will have the same range of swimming speeds.

1 Measured ordinal data is collected (go to box 2).

2 Two variables are correlated (go to box 4).

3 We are testing for a trend (go to box 6).

4 The variables are likely to follow a normal distribution (go to box 9).

5 Use the correlation coefficient test.

REMEMBER: Tests will give you a probability of the data distribution being due to chance alone. In biology we look for a probability of 5% or less to allow us to say that the data differ from chance significantly. The 'critical values' of any test are therefore the figures that will give 5% probability.

SUMMARY QUESTIONS

Suggest suitable tests for each of the following situations. (For questions 2, 3, 7, 8, 9, 10 and 11 you should assume that the data is normally distributed.)

1 Abundance of celandines in beech (5 samples) and oak (8 samples) woodland.

2 Width of ivy leaves in shaded and light areas, 30 samples from each area.

3 Number of beetles in two populations, 15 samples per population.

4 Number of daisies in four soil humidities, total number of daisies per site.

5 Length of time to collect 15 cm^3 O_2 when varying [HCO_3^-] in *Elodea* bubblers. Ten [HCO_3^-] tested.

6 Number of woodlice in the four quarters of a choice chamber.

7 Number of lymphocytes per mm^3 blood in people before and after infection with common colds, 25 volunteers tested.

8 Aperture width compared with shell length in dog whelks, 30 animals measured.

9 Aperture width to shell length mean ratio in dog whelks on two beaches, 30 animals measured per beach.

10 Ratio of egg mass to body mass in mackerel, 40 fish tested.

11 Recovery time after a fixed period of exercise before and after training, 20 volunteers tested.

12 The number of ladybirds on nettles compared with the number on docks (5 samples from each plant type).

13 Species diversity in rock pools of different volumes, 28 rock pools measured.

14 Possible correlation between presence of flat periwinkles and bladder wrack, 56 quadrats observed.

15 Abundance of flat periwinkles compared with abundance of bladder wrack, 56 quadrats observed.

16 Ratio of phenotypes in a genetic dihybrid cross.

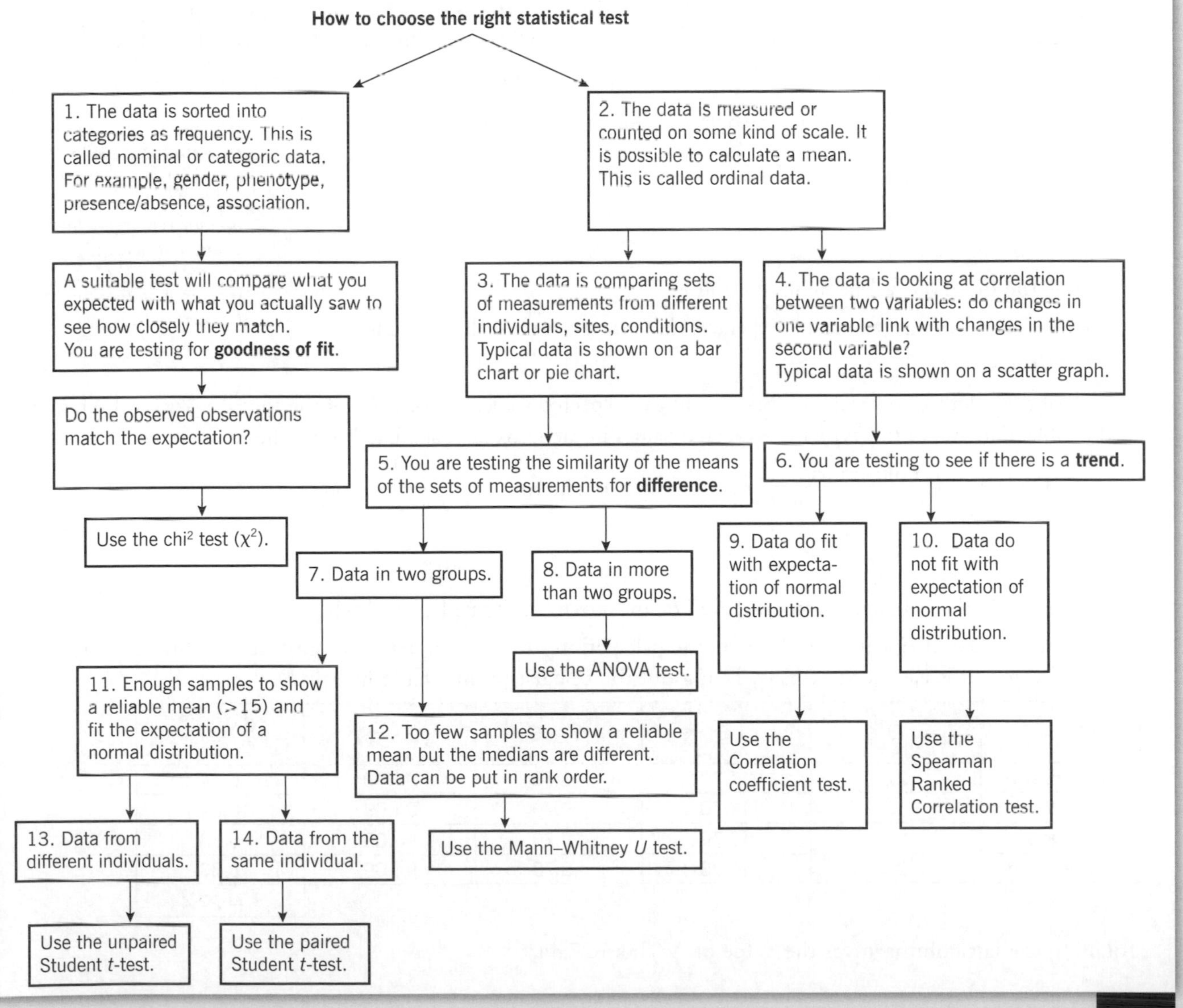

Tests of association

Using tests of association

When a biologist carries out sampling it may appear that there is a link between different observations. For example, on rocky shores the flat periwinkle *Littorina littoralis* is common among the fronds of wrack *Fucus vesiculosus*, so where one is found so is the other. An association analysis test is a way of demonstrating the closeness of the apparent link. If the test results in a significant association then the biologist will have good reason to explore further to try and identify the cause of the link.

WORKED EXAMPLE

The procedure begins with the gathering of data using a suitable random technique. This is often done by placing quadrats at random within the sample area. It is vital that the sampling is done at random. The data collected is written into a *contingency* table like the example below.

Observed occurrences (*O*)	*Fucus vesiculosus* present	*Fucus vesiculosus* absent	Totals
Littorina littoralis present	27	2	29
Littorina littoralis absent	7	21	28
Totals	34	23	57

Expected occurrences (*E*)	*Fucus vesiculosus* present	*Fucus vesiculosus* absent	Totals
Littorina littoralis present	(34 × 29) ÷ 57 = 17.30	(23 × 29) ÷ 57 = 11.70	29
Littorina littoralis absent	(34 × 28) ÷ 57 = 16.70	(23 × 28) ÷ 57 = 11.30	28
Totals	34	23	57

To calculate the significance of any association the chi^2 test is used.

The formula for this is $\chi^2 = \sum \frac{(O - E)^2}{E}$

In the equation:

χ^2 = chi squared

O = the observed values when sampling

E = the values that would be expected if the data were just due to chance

$\sum$ = the sum of.

> **REMEMBER:** We look for a probability of 5% or less to allow us to say that the data differ from chance significantly. The 'critical values' of any test are therefore the figures that will give 5% probability.

Step 1 of the calculation involves working out the expected values, E, for each of the four possible combinations in the table. For data of this type the expected values in each box are calculated using the formula:

expected value $E = \frac{\text{column total} \times \text{row total}}{\text{grand total}}$, so for the top left box (both present)

$$E = \frac{34 \times 29}{57} = 17.30$$

In this example the calculated expected values are shown in the right hand table.

Now that the O and E values are known, the calculation can be completed by substituting into the formula. To avoid errors it is easy to do this by laying out the calculation in a table like this:

	O	*E*	*O* − *E*	(*O* − *E*)^2	$\frac{(O - E)^2}{E}$
Both present	27	17.30	9.70	94.09	5.44
Only *Littorina*	2	11.70	−9.70	94.09	8.04
Only *Fucus*	7	16.70	−9.70	94.09	5.63
Both absent	21	11.30	9.70	94.09	8.33
					χ^2 = 27.44

Totalling the last column gives the value of χ^2, here 27.44

Next calculate the number of degrees of freedom, which for a contingency table is (the number of rows − 1) × (the number of columns − 1).
This gives (2 − 1) × (2 − 1) = 1 degree of freedom in the data.

Finally, look up the calculated value of χ^2 in the χ^2 table to find the probability of the observed data being just due to chance.

In the table 27.44 at one degree of freedom corresponds with a probability (p) of <0.001 or <0.1% that the data are simply due to chance.

$p < 0.001$

1 degree of freedom uses this row

27.44 lies to the right of the range available

df	*p* values								df
	0.99	0.95	0.90	0.50	0.10	0.05	0.01	0.001	
1	0.0016	0.0039	0.016	0.46	2.71	3.84	6.63	10.83	1
2	0.02	0.10	0.21	1.39	4.60	5.99	9.21	13.82	2
3	0.12	0.35	0.58	2.37	6.25	7.81	11.34	16.27	3
4	0.30	0.71	1.06	3.36	7.78	9.49	13.28	18.46	4
5	0.55	1.14	1.61	4.35	9.24	11.07	15.09	20.52	5
6	0.87	1.64	2.20	5.35	10.64	12.59	16.81	22.46	6
7	1.24	2.17	2.83	6.35	12.02	14.07	18.48	24.32	7
8	1.65	2.73	3.49	7.34	13.36	15.51	20.09	26.12	8
9	2.09	3.32	4.17	8.34	14.68	16.92	21.67	27.88	9
10	2.56	3.94	4.86	9.34	15.99	18.31	23.21	29.59	10
11	3.05	4.58	5.58	10.34	17.28	19.68	24.72	31.26	11
12	3.57	5.23	6.30	11.34	18.55	21.03	26.22	32.91	12
13	4.11	5.89	7.04	12.34	19.81	22.36	27.69	34.53	13
14	4.66	6.57	7.79	13.34	21.06	23.68	29.14	36.12	14
15	5.23	7.26	8.55	14.34	22.31	25.00	30.58	37.70	15
16	5.81	7.96	9.31	15.34	23.54	26.30	32.00	39.29	16
17	6.41	8.67	10.08	16.34	24.77	27.59	33.41	40.75	17
18	7.02	9.39	10.86	17.34	25.99	28.87	34.80	42.31	18
19	7.63	10.12	11.65	18.34	27.20	30.14	36.19	43.82	19
20	8.26	10.85	12.44	19.34	28.41	31.41	37.57	45.32	20
21	8.90	11.59	13.24	20.34	29.62	32.67	38.93	46.80	21
22	9.54	12.34	14.04	21.34	30.81	33.92	40.29	48.27	22
23	10.20	13.09	14.85	22.34	32.01	35.17	41.64	49.73	23
24	10.86	13.85	15.66	23.34	33.20	36.42	42.98	51.18	24
25	11.52	14.61	16.47	24.34	34.38	37.65	44.31	52.62	25
26	12.20	15.38	17.29	25.34	35.56	38.88	45.64	54.05	26
27	12.88	16.15	18.11	26.34	36.74	40.11	46.96	55.48	27
28	13.56	16.93	18.94	27.34	37.92	41.34	48.28	56.89	28
29	14.26	17.71	19.77	28.34	39.09	42.56	49.59	58.30	29
30	14.95	18.49	20.60	29.34	40.26	43.77	50.89	59.70	30
40	22.16	26.51	29.05	39.34	51.81	55.76	63.69	73.40	40
60	37.48	43.19	46.46	59.33	74.40	79.08	88.38	99.61	60
80	53.54	60.39	64.28	79.33	96.58	101.88	112.33	124.84	80
100	70.06	77.93	82.36	99.33	118.50	124.34	135.81	149.45	100

This column shows the critical values at 5%. Chi² this big or more gives significance.

Table of values of chi²

This means that the observations are 99.9% certain not to be chance, so some factor must be causing the inequality. Looking at the table again it is clear that the two species are most often together, which suggests a positive association. Since the snail eats the alga, this is not unexpected!

REMEMBER: This test needs representative samples that have been collected at random. Beware of bias due to the time when the samples are taken!

SUMMARY QUESTION

1 Use chi² to determine the strength of association, if any, in the following examples.

	Species A	Species B	Number of quadrats with species A only	Number of quadrats with species A and B	Number of quadrats with species B only	Number of quadrats with neither species
a	dog whelk	barnacle	7	19	5	9
b	ladybird	aphid	2	21	9	12
c	dock	nettle	8	10	7	8

Correlation coefficient

Calculating a correlation coefficient

When sampling collects data from two variables it is possible to use a calculation to determine whether the two variables correlate in any way. For example, does the leaf surface area of a plant increase as plant height increases? The variables need to be plotted on a scatter graph, which will indicate possible relationships that can then be tested.

WORKED EXAMPLE

Neil sampled conkers from a horse chestnut tree. When he plotted his data on a scatter graph a pattern emerged. The pattern appeared to show that as conker seed mass increased the number of wrinkles on the seeds decreased. The wrinkling appeared to be negatively correlated with mass. How strong was this correlation?

The formula for correlation coefficient is $r = \dfrac{\sum[(x - x) \times (y - y)]}{\sqrt{\sum(x - x)^2 \times \sum(y - y)^2}}$

In the equation:

x = the values of the first variable

y = the values of the second variable

$\sum$ = the sum of

$\bar{x}$ = the average of the values of the first variable

$\bar{y}$ = the average of the values of the second variable

r = the correlation coefficient.

Use of a table helps the calculation.

Seed mass/g = x	Number of wrinkles per seed = y	$x - \bar{x}$	$y - \bar{y}$	$(x - \bar{x}) \times (y - \bar{y})$	$(x - \bar{x})^2$	$(y - \bar{y})^2$
12	1	5	−14	−70	25	196
10	3	3	−12	−36	9	144
8	8	1	−7	−7	1	49
6	15	−1	0	0	1	0
4	27	−3	12	−36	9	144
2	36	−5	21	−105	25	441
$\sum x = 42$	$\sum y = 90$			$\sum = -254$	$\sum = 70$	$\sum = 974$
$\bar{x} = 42 \div 6 = 7$	$\bar{y} = 90 \div 6 = 15$					

Now substitute the values from the table into the equation to find r.

$$r = \frac{\sum(x - \bar{x}) \times (y - \bar{y})}{\sqrt{\sum(x - \bar{x})^2 \times \sum(y - \bar{y})^2}}$$

$$r = \frac{-254}{\sqrt{(70 \times 974)}} = \frac{-254}{\sqrt{68\,180}} = \frac{-254}{261} = -0.97$$

Note that the sign is not changed during the steps of the calculation. The value of r varies between two limits. Perfect positive correlation gives a value of $r = 1$, while perfect negative correlation gives a value of $r = -1$. The nearer the value of r is to zero, the lower the strength of any correlation. In this case the correlation is negative, as wrinkle number decreases as seed mass increases.

To judge the statistical strength of the correlation the calculated value needs to be looked up in a table.

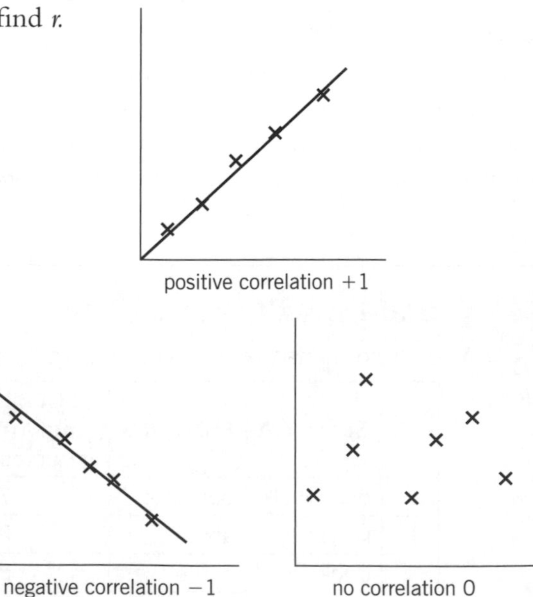

positive correlation +1

negative correlation −1

no correlation 0

In this test the number of degrees of freedom (df) = the number of values for the two variables (n) – 2.

In this example df = 12 – 2 = 10.

At 10 degrees of freedom a value of 0.97 gives a p value of <0.001 or <0.1% of the observed data being due only to chance. Thus the correlation is 99.9% certain.

df	*p* values			
	0.1	0.05	0.01	0.001
1	0.9877	0.9969	0.9999	1.0000
2	0.9000	0.9500	0.9900	0.9990
3	0.8054	0.8783	0.9587	0.9912
4	0.7293	0.8114	0.9172	0.9741
5	0.6694	0.7545	0.8745	0.9507
6	0.6215	0.7067	0.8343	0.9249
7	0.5822	0.6664	0.7977	0.8982
8	0.5494	0.6319	0.7646	0.8721
9	0.5214	0.6021	0.7348	0.8471
10	0.4973	0.5760	0.7079	0.8233
11	0.4762	0.5529	0.6835	0.8010
12	0.4575	0.5324	0.6614	0.7800
13	0.4409	0.5139	0.6411	0.7603
14	0.4259	0.4973	0.6226	0.7420
15	0.4124	0.4821	0.6055	0.7246
16	0.4000	0.4683	0.5897	0.7084
17	0.3887	0.4555	0.5751	0.6932
18	0.3783	0.4438	0.5614	0.6787
19	0.3687	0.4329	0.5487	0.6652
20	0.3598	0.4227	0.5368	0.6524

Probability is < 0.001

10 degrees of freedom uses this row

0.97 lies to the right of the range given

Table of values of *r*, correlation coefficient

SUMMARY QUESTION

1 Calculate correlation coefficients to check for correlation between these three sets of data. Remember to plot scatter graphs first.

Thorns on brambles		Alcoholic liver disease		Trout skin spots	
Stem diameter/mm	Number of thorns per 10 cm length	Alcohol consumption/units per week	Incidence of liver disease/cases per 100 000	Body length/cm	Number of spots per cm^2
2	5	40–44	91	40–45	18
3	17	36–40	73	35–40	22
4	11	32–36	61	30–35	15
5	11	28–32	48	25–30	16
6	14	24–28	12	20–25	9
7	31	20–24	9	15–20	21
8	22	16–20	2	10–15	12
9	13	12–16	1		

Ranked correlation coefficient

Using a ranked correlation coefficient

If a biologist is trying to establish a correlative relationship between two variables that may be assumed to have non normal distribution they may use the Spearman Ranked Correlation test. The test will give an indication of positive or negative correlation and can also be used to judge the statistical strength of any correlation demonstrated. Plotting the two variables on a scatter graph will give a pictorial indication of the relationship, but the test adds the value of a numerical expression of certainty.

WORKED EXAMPLE

Caro wants to find out whether the abundance of daisies is linked with the abundance of moss in a lawn.

First she uses random quadrat samples A to G to measure the percentage cover of each type of plant. She records the values in the table in columns 2 and 4.

The next step involves ranking the values of each variable from smallest to largest. If two values are identical then they should be ranked equally. The ranks are placed in the table in columns 3 and 5.

The formula for the Spearman Rank Correlation r_s is:

$$r_s = 1 - \frac{6\sum d^2}{n(n^2 - 1)}$$

d is the difference between the ranks at each site (column 6; d^2 in column 7)

n is the number of samples

$\sum$ is the sum of.

1	2	3	4	5	6	7
Sample site	Moss % cover		Daisy % cover			
	Value	Rank	Value	Rank	Difference in ranks d	d^2
A	0.5	1	97	7	−6	36
B	2.2	2	88	6	−4	16
C	2.7	3	88	5	−2	4
D	4.4	4.5	56	4	0.5	0.25
E	4.4	4.5	49	3	1.5	2.25
F	8.5	6	21	2	4	16
G	10.2	7	17	1	6	36
						$\sum d^2 = 110.5$

Substituting the values from the table into the equation gives:

$$r_s = 1 - \frac{6\sum d^2}{n(n^2 - 1)} \qquad r_s = 1 - \frac{6 \times 110.5}{7 \times 48} = 1 - \frac{663}{336} = -0.97$$

The negative value indicates negative correlation, i.e. the mosses decline as the daisies increase. This value can be looked up in a statistical table of r_s, which on this occasion shows the significance of this correlation to be <0.005 or $<0.5\%$ due to chance, or 99.5% certain to be due to some other factor.

Number of pairs of values (n)	p values				
	0.10	0.05	0.025	0.01	0.005
4	1.0000	1.0000	1.0000	1.0000	1.0000
5	0.7000	0.9000	0.9000	1.0000	1.0000
6	0.6571	0.7714	0.8286	0.9429	0.9429
7	0.5714	0.6786	0.7857	0.8571	0.8929
8	0.5476	0.6429	0.7381	0.8095	0.8571
9	0.4833	0.6000	0.6833	0.7667	0.8167
10	0.4424	0.5636	0.6485	0.7333	0.7818
11	0.4182	0.5273	0.6091	0.7000	0.7545
12	0.3986	0.5035	0.5874	0.6713	0.7273
13	0.3791	0.4780	0.5604	0.6484	0.6978
14	0.3670	0.4593	0.5385	0.6220	0.6747
15	0.3500	0.4429	0.5179	0.6000	0.6536
16	0.3382	0.4265	0.5029	0.5824	0.6324
17	0.3271	0.4124	0.4821	0.5577	0.6055
18	0.3170	0.4000	0.4683	0.5425	0.5897
19	0.3077	0.3887	0.4555	0.5285	0.5751
20	0.2992	0.3783	0.4438	0.5155	0.5614

7 pairs of values use this row

Likelihood of correlation being due to chance is <0.005

0.97 lies to the right of the range given

If r_s appears in or to the right of this column then this means there is less than a 5% chance of the results being coincidental.

Table of values of r_s, Spearman Ranked Correlation

REMEMBER: Use this test for data that is not expected to show a normal distribution e.g. one variable is a linear scale.

SUMMARY QUESTION

1 Calculate the Spearman Rank Correlation for the three data sets below. How strong is the correlation in each case?

Fucus on a rocky shore	
Distance up shore/m	% cover by *Fucus*
3	100
6	94
9	86
12	62
15	63
18	41
21	22
24	17

Leaf size of a plant	
Light intensity at growth site/lux	Leaf surface area/cm^2
200	40
300	23
400	20
500	19
600	36
700	45
800	55
900	54

Elodea photosynthesis	
$[HCO_3^-]$/%	Volume of oxygen produced/ $mm^3 min^{-1}$
0.05	3.0
0.10	3.0
0.15	3.0
0.20	3.0
0.25	7.4
0.30	7.4
0.35	10.2
0.40	11.4

Tests of difference: unpaired *t*-tests

Using tests of difference: unpaired *t*-tests

Data that are collected from two groups but which are not paired measurements of the same individuals can be tested using the Student *t*-test. There must be enough data to calculate a reliable mean and it must be normally distributed. The number in each sample does not need to be the same.

WORKED EXAMPLE

Alistair measured limpet heights at two sites to find out whether there was any difference due to aspect. One site faced east and the other west. He measured 28 animals at each site.

The null hypothesis is that there will be no difference between the sites.

	Site 1 (east)	Site 2 (west)
n	28	28
Mean limpet diameter/mm	35.64	37.36
Variance (s^2) (= the square of standard deviation)	77.17	74.40

First, calculate the value of t, using this formula:

$$t = \frac{\bar{x}_1 - \bar{x}_2}{\sqrt{\frac{s_1^2}{n_1} + \frac{s_2^2}{n_2}}}$$

Where s_1^2 and s_2^2 are the variances at each site, n_1 and n_2 are the numbers sampled at each site and $\bar{x}_1$ and $\bar{x}_2$ are the means for each site.

So substituting in the values from the table:

$$t = \frac{35.64 - 37.36}{\sqrt{\frac{77.17}{28} + \frac{74.4}{28}}} = \frac{-1.72}{2.33}$$

(ignore the − sign) $= 0.74$

Next, calculate the degrees of freedom. For an unpaired test this is: $(n_1 + n_2) - 2 = 54$

Degree of freedom (df)	*p* values			
	0.10	0.05	0.01	0.001
1	6.31	12.71	63.66	636.60
2	2.92	4.30	9.92	31.60
3	2.35	3.18	5.84	12.92
4	2.13	2.78	4.60	8.61
5	2.02	2.57	4.03	6.87
6	1.94	2.45	3.71	5.96
7	1.89	2.36	3.50	5.41
8	1.86	2.31	3.36	5.04
9	1.83	2.26	3.25	4.78
10	1.81	2.23	3.17	4.59
12	1.78	2.18	3.05	4.32
14	1.76	2.15	2.98	4.14
16	1.75	2.12	2.92	4.02
18	1.73	2.10	2.88	3.92
20	1.72	2.09	2.85	3.85
22	1.72	2.08	2.82	3.79
24	1.71	2.06	2.80	3.74
26	1.71	2.06	2.78	3.71
28	1.70	2.05	2.76	3.67
30	1.70	2.04	2.75	3.65
40	1.68	2.02	2.70	3.55
60	1.67	2.00	2.66	3.46
120	1.66	1.98	2.62	3.37
∞	1.64	1.96	2.58	3.29

Probability value is >0.10 or >10%

t value 0.74 lies to the left of the range given

54 degrees of freedom uses this row

Table of values of *t*

In this case the probability of the difference between the means being due to chance alone is more than 10%, so the difference is not significant. The null hypothesis is accepted.

REMEMBER: This test compares means. At least 15 samples are required, which must be collected at random.

For the data to be considered significantly different from chance alone, the probability must be 5% or less.

SUMMARY QUESTIONS

1 Mike was testing the effectiveness of glass cloches compared with plastic cloches. Using the same strain of seed, he grew 15 seeds under plastic cloches and 15 seeds under glass cloches. He measured the amount of time they took to flower. His results are below. Use a *t*-test to see if there is any significant difference between the two types of cloche.

	A plastic	B glass
Number (n)	15	15
Mean of time taken / days	99	100
Variance (s^2)	3.2	2.8

2 Caroline bought some apples from a shop to carry out an experiment. Their masses (g) were

180, 170, 160, 166, 177, 198, 199, 200, 167, 300, 169, 191, 194, 203, 201

A separate batch was bought from another shop. Their masses were:

170, 124, 135, 154, 167, 156, 134, 132, 133, 154, 156, 145, 160, 136, 132

Using a *t*-test, decide if these two sets of apples are significantly different, and if Caroline should use them in the same experiment.

3 One field of oats had been treated with fertiliser, a second field had not been treated. The mean heights of the plants in each field were found. Using a *t*-test, decide if the fertiliser had a significant effect on the oat growth.

	Fertiliser	No fertiliser
Number (n)	289	265
Mean height/m	0.96	0.85
Standard deviation (s)	0.03	0.10

Differences in data: paired *t*-tests

Identifying differences in data: paired *t*-tests

If data is collected from the same individuals so that the samples are in pairs, then the paired *t*-test may be used. There must be the same number of measurements in each sample and enough measurements must be taken to allow a reliable mean to be calculated. At least 15 measurements are recommended. The test compares the means to see if the samples differ significantly. The data tested will show a normal distribution.

WORKED EXAMPLE

A biology student investigated the number of red pustule galls (caused by a mite called *Aceria macrorhynchus*) found on sycamore leaves. The student thought that the galls would be more abundant on the upper surfaces of the leaves. The student examined 20 leaves (so $n = 20$) and counted the total number of galls on each side.

Leaf	Number of galls – top surface	Number of galls – lower surface	Difference d	Difference squared d^2
1	78	45	33	1089
2	82	33	49	2401
3	56	37	19	361
4	67	42	25	625
5	69	23	46	2116
6	78	14	64	4096
7	71	32	39	1521
8	45	21	24	576
9	69	18	51	2601
10	67	18	49	2401
11	91	31	60	3600
12	83	26	57	3249
13	78	21	57	3249
14	67	19	48	2304
15	76	12	64	4096
16	78	29	49	2401
17	90	43	47	2209
18	83	36	47	2209
19	76	15	61	3721
20	89	22	67	4489
		Totals	956	49 314

Step 1: find the difference between each pair of readings, d. This is shown in column 4 of the table.

Step 2: calculate the mean of the differences by using $\bar{d} = \frac{\sum d}{n}$

$\sum d$ is the total of all the differences, here 956, so the mean of the differences $\bar{d}$ is $\frac{956}{20} = 47.8$

Step 3: calculate the standard deviation of the differences using $s_d = \sqrt{\frac{\sum d^2 - n\bar{d}^2}{n - 1}}$

So $s_d = \sqrt{\frac{49\,314 - 20 \times 47.8^2}{19}} = 13.80$

Step 4: calculate t using $t = \frac{\bar{d}\sqrt{n}}{s_d}$ So $t = \frac{47.8\sqrt{20}}{13.80} = 15.49$

Step 5: find the number of degrees of freedom, which is the number of pairs − 1, i.e. 19 in this case.

Step 6: look up the t value to find the probability of the data being only due to chance.

Degree of freedom (df)	p values			
	0.10	0.05	0.01	0.001
1	6.31	12.71	63.66	636.60
2	2.92	4.30	9.92	31.60
3	2.35	3.18	5.84	12.92
4	2.13	2.78	4.60	8.61
5	2.02	2.57	4.03	6.87
6	1.94	2.45	3.71	5.96
7	1.89	2.36	3.50	5.41
8	1.86	2.31	3.36	5.04
9	1.83	2.26	3.25	4.78
10	1.81	2.23	3.17	4.59
12	1.78	2.18	3.05	4.32
14	1.76	2.15	2.98	4.14
16	1.75	2.12	2.92	4.02
18	1.73	2.10	2.88	3.92
20	1.72	2.09	2.85	3.85
22	1.72	2.08	2.82	3.79
24	1.71	2.06	2.80	3.74
26	1.71	2.06	2.78	3.71
28	1.70	2.05	2.76	3.67
30	1.70	2.04	2.75	3.65
40	1.68	2.02	2.70	3.55
60	1.67	2.00	2.66	3.46
120	1.66	1.98	2.62	3.37
∝	1.64	1.96	2.58	3.29

Probability value is <0.001 or <0.1%

t value 15.49 lies to the right of the range given

19 degrees of freedom uses this row

Table of values of *t*

In this case the data show a clear improbability of chance, so a difference between the galls on the surface is genuine and must be due to a reason other than chance. The null hypothesis is rejected.

REMEMBER: This test compares means. To use this test the data must be in pairs, e.g. from the same individual and at least 15 samples are required.

For the data to be considered significantly different from chance alone, the probability must be 5% or less.

SUMMARY QUESTIONS

1 In a test comparing 28 paired samples the value of *t* was found to be 2.46. What is the probability of the samples differing just by chance?

2 Three populations of pond snails were tested to find whether the rate of shell growth could be increased by the addition of calcium salts to the water. Population A (14 individuals) gave a value of $t = 1.87$; population B (26 individuals) gave a value of $t = 3.68$; population C (16 individuals) gave a value of $t = 2.01$. It was concluded that only one population showed a significant growth response to the added calcium salts. Explain which one this was.

3 Calculate values of *t* for the following examples.

Heart rates of different people treated with a new drug		Hormone levels in 15 adult ibis/ng cm^{-3}	
Heart rate before drug/ beats per minute	Heart rate after drug/ beats per minute	Winter	Summer
87	76	0.9	2.5
92	78	0.8	2.1
89	67	0.4	2.4
89	79	0.8	2.3
71	67	0.6	1.9
89	68	0.5	2.1
81	76	0.6	1.3
81	70	0.3	1.6
83	74	0.6	1.4
79	69	0.6	1.9
82	67	0.9	2.3
71	56	0.4	2.1
89	70	0.7	1.0
88	72	0.9	2.9
85	68	0.6	2.7

The Mann–Whitney *U* test

Using the Mann–Whitney *U* test

When small quantities of data are collected it is not possible to be confident in the reliability of the mean or to make a clear judgement about whether the data fit a normal distribution. However, sites may be compared to see how well the data overlap using the Mann–Whitney *U* test, as long as the data can be ranked. Like the *t*-test, the Mann–Whitney *U* test tests for differences. This test is also better than the *t*-test at allowing for errors due to outliers in the data.

WORKED EXAMPLE

Bridget wanted to compare the density of a species of lichen on two rock types. She collected some data in a table.

% cover of lichen using 10 cm × 10 cm quadrats								
Quadrat	1	2	3	4	5	6	7	8
Site 1 Limestone	12	10	21	25	8	7	13	26
Site 2 Sandstone	29	24	27	25	15	29	16	23

Step 1: calculate the median values for each site. This provides a quick check on whether to proceed – remember the data are insufficient to be confident about a mean value. If the median values are different then proceed with the test.

The median for limestone is 12.5 and for sandstone 24.5.

Step 2: arrange all the data in order and rank the values, giving the smallest value the lowest rank. If two values are the same they get the average of the two ranks available.

Site 1 Limestone	7	8	10	12	13			21				25	26			
Site 2 Sandstone						15	16		23	24	25			27	29	29
Rank	1	2	3	4	5	6	7	8	9	10	11.5	11.5	13	14	15.5	15.5

Step 3: find the sum of the ranks, $\sum R$, for each set of data.

$$\sum R_{\text{limestone}} = 1 + 2 + 3 + 4 + 5 + 8 + 11.5 + 13 = 47.5$$

$$\sum R_{\text{sandstone}} = 6 + 7 + 9 + 10 + 11.5 + 14 + 15.5 + 15.5 = 88.5$$

Step 4: calculate U_1 and U_2 using the formulae:

$$U_1 = n_1 \times n_2 + \tfrac{1}{2}\, n_2(n_2 + 1) - \sum R_{\text{sandstone}}$$

$$U_2 = n_1 \times n_2 + \tfrac{1}{2}\, n_1(n_1 + 1) - \sum R_{\text{limestone}}$$

n_1 and n_2 are the numbers of quadrats at sites 1 and 2.

$$U_1 = 8 \times 8 + \tfrac{1}{2}\, 8(8 + 1) - 88.5 = 11.5$$

$$U_2 = 8 \times 8 + \tfrac{1}{2}\, 8(8 + 1) - 47.5 = 52.5$$

Step 5: compare the smaller of the two calculated *U* values with the critical value table. If the calculated *U* value is less than or equal to the critical value then there is a significant difference between the data and a more thorough collection of data followed by a *t*-test may be warranted.

The critical U value is 13.

Bridget had 8 samples in each site.

Mann–Whitney U test critical values at $p = 0.05$ (5%)																
Size of the smaller sample (n_1)	Size of the larger sample (n_2)															
	5	6	7	8	9	10	11	12	13	14	15	16	17	18	19	20
3	0	1	1	2	2	3	3	4	4	5	5	6	6	7	7	8
4	1	2	3	4	4	5	6	7	8	9	10	11	11	12	13	14
5	2	3	5	6	7	8	9	11	12	13	14	15	17	18	19	20
6		5	6	8	10	11	13	14	16	17	19	21	22	24	25	27
7			8	10	12	14	16	18	20	22	24	26	28	30	32	34
8				13	15	17	19	22	24	26	29	31	34	36	38	41
9					17	20	23	26	28	31	34	37	39	42	45	48
10						23	26	29	33	36	39	42	45	48	52	55
11							30	33	37	40	44	47	51	55	58	62
12								37	41	45	49	53	57	61	65	69
13									45	50	54	59	63	67	72	76
14										55	59	64	67	74	78	83
15											64	70	75	80	85	90
16												75	81	86	92	98
17													87	93	99	105
18														99	106	112
19															113	119
20																127

Table of Mann–Whitney U test critical values

In this example the smaller value of U was 11.5. Since this was less than the critical U value of 13 the conclusion is that there is significant difference between the two sites.

REMEMBER: Use this test when small sample sizes are available and the medians differ.

The test can be used even when the number of samples at the two sites is different.

SUMMARY QUESTIONS

Test the following sets of data to determine whether the samples differ significantly.

1

Number of sundew plants in two areas										
Quadrat	1	2	3	4	5	6	7	8	9	10
Site 1 (wet)	6	9	12	4	2	17	9	11	12	8
Site 2 (dry)	2	4	1	1	3	4	1	3	2	1

2

Number of yellow banded snails in hedges and woodland										
Quadrat	1	2	3	4	5	6	7	8	9	10
Site 1 woodland	2	1	7	3	0	1	1	0	1	4
Site 2 grassland	5	8	3	6	9	9	4	6	3	9

3 Determine whether the following pairs of samples differ significantly

Site 1	Site 2	Value of U_1	Value of U_2
4 samples of limpets	16 samples of limpets	8.3	14.6
14 samples of flour beetles	14 samples of flour beetles	87.0	72.0
12 samples of ivy leaves	8 samples of ivy leaves	18.0	8.5

Chi-squared goodness of fit test in genetics

Using the chi-squared goodness of fit test in genetics

Sometimes, particularly in genetics, the results of crosses are expected to fit a predicted ratio. For example, in a heterozygous monohybrid cross the number of each phenotype should match a 3 : 1 ratio, while a heterozygous dihybrid cross should fit a 9 : 3 : 3 : 1 ratio. A chi^2 goodness of fit test may be used to check the agreement between what was observed and what was expected.

WORKED EXAMPLE

Alistair carried out crosses to investigate the inheritance of stem height in peas. Pea plants can be tall (dominant) or short (recessive). All the parent plants he used were heterozygous, so he expected to see a 3 : 1 ratio among the offspring. Alistair grew 764 F1 plants, of which 590 turned out tall and 174 short. These were his observed numbers. According to theory he expected a 3 : 1 ratio, i.e. $764 \times 0.75 = 573$ tall and $764 \times 0.25 = 191$ short plants because all alleles are inherited by chance. How closely did the observations match the predictions?

The formula for the chi^2 (χ^2) test is $\chi^2 = \sum \frac{(O - E)^2}{E}$. It is easiest to lay out the calculation in a table.

	Observed (O)	Expected (E)	($O - E$)	$(O - E)^2$	$\frac{(O - E)^2}{E}$
Tall pea plants	590	573	17	289	0.50
Dwarf pea plants	174	191	−17	289	1.51
				Total:	$\chi^2 = 2.01$

The significance of the χ^2 value can be found by looking in statistical tables. First, work out the degrees of freedom, which is the number of categories minus 1. Here there are two categories (tall or short), so there is 1 degree of freedom.

The critical value of χ^2 for one degree of freedom at the $p = 5\%$ level is 3.84. The calculated value of 2.01 is less than 3.84 and gives a p value of between 0.1 and 0.5 (10% and 50%) of the data collected being just chance. In this case Alistair should accept the null hypothesis; there is no evidence of a significant difference between the observed and expected numbers.

SUMMARY QUESTIONS

1 Use the chi^2 method to test whether the following data from heterozygous dihybrid crosses fit a 9 : 3 : 3 : 1 ratio.

  a In peas, 315 tall, round seeded : 101 tall, wrinkled seeded : 108 short, round seeded : 32 short, wrinkled seeded.

  b In fruit flies, 999 red eyes, straight wings: 462 red eyes, vestigial wings : 350 white eyes, straight wings : 111 white eyed, vestigial wings.

2 When two pink *Antirrhinum* plants produced seed, the 177 F1 plants grew in the ratio 47 red : 86 pink : 44 white. In this type of co-dominant inheritance a 1 : 2 : 1 ratio was expected.

  Use chi^2 to test how well the results fitted with expectation.

df	p values								df
	0.99	0.95	0.90	0.50	0.10	0.05	0.01	0.001	
1	0.0016	0.0039	0.016	0.46	2.71	3.84	6.63	10.83	1
2	0.02	0.10	0.21	1.39	4.60	5.99	9.21	13.82	2
3	0.12	0.35	0.58	2.37	6.25	7.81	11.34	16.27	3
4	0.30	0.71	1.06	3.36	7.78	9.49	13.28	18.46	4
5	0.55	1.14	1.61	4.35	9.24	11.07	15.09	20.52	5
6	0.87	1.64	2.20	5.35	10.64	12.59	16.81	22.46	6
7	1.24	2.17	2.83	6.35	12.02	14.07	18.48	24.32	7
8	1.65	2.73	3.49	7.34	13.36	15.51	20.09	26.12	8
9	2.09	3.32	4.17	8.34	14.68	16.92	21.67	27.88	9
10	2.56	3.94	4.86	9.34	15.99	18.31	23.21	29.59	10
11	3.05	4.58	5.58	10.34	17.28	19.68	24.72	31.26	11
12	3.57	5.23	6.30	11.34	18.55	21.03	26.22	32.91	12
13	4.11	5.89	7.04	12.34	19.81	22.36	27.69	34.53	13
14	4.66	6.57	7.79	13.34	21.06	23.68	29.14	36.12	14
15	5.23	7.26	8.55	14.34	22.31	25.00	30.58	37.70	15
16	5.81	7.96	9.31	15.34	23.54	26.30	32.00	39.29	16
17	6.41	8.67	10.08	16.34	24.77	27.59	33.41	40.75	17
18	7.02	9.39	10.86	17.34	25.99	28.87	34.80	42.31	18
19	7.63	10.12	11.65	18.34	27.20	30.14	36.19	43.82	19
20	8.26	10.85	12.44	19.34	28.41	31.41	37.57	45.32	20
21	8.90	11.59	13.24	20.34	29.62	32.67	38.93	46.80	21
22	9.54	12.34	14.04	21.34	30.81	33.92	40.29	48.27	22
23	10.20	13.09	14.85	22.34	32.01	35.17	41.64	49.73	23
24	10.86	13.85	15.66	23.34	33.20	36.42	42.98	51.18	24
25	11.52	14.61	16.47	24.34	34.38	37.65	44.31	52.62	25
26	12.20	15.38	17.29	25.34	35.56	38.88	45.64	54.05	26
27	12.88	16.15	18.11	26.34	36.74	40.11	46.96	55.48	27
28	13.56	16.93	18.94	27.34	37.92	41.34	48.28	56.89	28
29	14.26	17.71	19.77	28.34	39.09	42.56	49.59	58.30	29
30	14.95	18.49	20.60	29.34	40.26	43.77	50.89	59.70	30
40	22.16	26.51	29.05	39.34	51.81	55.76	63.69	73.40	40
60	37.48	43.19	46.46	59.33	74.40	79.08	88.38	99.61	60
80	53.54	60.39	64.28	79.33	96.58	101.88	112.33	124.84	80
100	70.06	77.93	82.36	99.33	118.50	124.34	135.81	149.45	100

$0.1 < p < 0.5$

The calculated value lies between these values

Table of values of chi²

3 A character for leg length in dogs has a dominant allele for long legs. Short legged animals always turned out to be heterozygous. Heterozygous short legged animals were bred and the 36 offspring turned out in the ratio 34 tall : 2 short.

a Test these results against the expected 3 : 1 monohybrid heterozygous ratio using chi².

b The breeder suspected that the recessive gene was lethal in utero when homozygous.
What would be the expected numbers of tall and short dogs if this was true?

REMEMBER: The chi-squared test can only be used with data that are counts, not with measurements.

The Hardy–Weinberg equation

Applying the Hardy–Weinberg equation

The frequency of dominant and recessive alleles in a population will remain in equilibrium from generation to generation, provided that:

- the population is large
- mating is random
- no mutations occur
- all genotypes and phenotypes are equally fertile
- generations do not overlap
- there is no emigration/immigration.

Thus it is possible to apply a statistical analysis to the population in order to establish the gene frequencies of the dominant and recessive alleles and the frequency of carriers in a population. This is the Hardy–Weinberg equation.

WORKED EXAMPLE

In the equation p represents the dominant and q the recessive allele.

Consider the cross between 2 heterozygous parents, **Aa**:

Parent genotypes	Aa		Aa	
Parent gametes	A	a	A	a
F1 genotypes	AA	Aa	Aa	aa
Substitute p and q	$p \times p$	$p \times q$	$p \times q$	$q \times q$
To give	p^2	$2pq$		q^2
so	$p^2 + 2pq + q^2 = 1$ (i.e. 100%)			

Note the sum of all alleles is always 1, so $p + q = 1$

Example

Cystic fibrosis occurs in a population with a frequency of $\frac{1}{2000}$ births. What is the frequency of the carrier genotype (i.e. the heterozygote)?

First find the frequency of the sufferer (q^2), which is the double recessive.

Here it is $\frac{1}{2000} = 0.0005$

If $q^2 = 0.0005$ then $q = \sqrt{0.0005}$

So $q = 0.0224$

Since $p + q = 1$ we can now find p, which is $1 - 0.0224 = 0.9776$

Therefore $2pq = 2(0.9776 \times 0.0224) = 0.044$ (4.4%)

So 4.4% of people are carriers.

The frequency of the homozygous dominant must therefore be $1 - (0.044 + 0.0005) = 0.956$ or 95.6% of the population.

REMEMBER: Always begin by determining the value of p or q. It is usually easiest to find q, because q^2 represents all the double recessives, which are very easy to identify in the population.

SUMMARY QUESTIONS

1 Around 1 in 17 000 UK people have some form of albinism, a recessive condition.

- **a** What is the frequency of the dominant allele?
- **b** What is the incidence of carriers in the population?
- **c** If the UK population is taken to be 63 million, how many people will be carriers of an albino allele?

2 Roberts syndrome is a rare recessive autosomal prenatal and postnatal growth disorder. It has an incidence of approximately 1 out of every million births.

- **a** What is the frequency of the carrier genotype?
- **b** What percentage of the population do not have the recessive allele in their genotypes?

3 Sickle cell anaemia occurs in $\frac{1}{500}$ African–American births but has a prevalence of 2% in West Africa.

- **a** Determine the frequency of carriers in both the USA and West Africa.
- **b** Explain the difference between the prevalence of carriers in these related populations.

4 Huntington's disease is caused by an autosomal dominant allele. It has an incidence of around 7 per 100 000 of the population.

- **a** Determine the frequency of the dominant and recessive alleles.
- **b** What proportion of sufferers are likely to be heterozygous?

Common calculator functions

Introducing common calculator functions

Biologists need to be able to use calculators confidently, but the types of calculations they perform are often relatively simple. Many calculators are available, but it is sensible to choose a simple one as the range of functions available on the very expensive calculators may be more than you require for biology.

WORKED EXAMPLES

Add/subtract/divide/multiply

Use keys [+], [−], [÷], [×].

For example, 67 × 7 enter 67, press [×], press 7, press [=]. You will get 469.

You can quickly add a string of numbers by using the [M+] key.

For example, add 3, 4, 5 and 6. Enter 3, press [M+], enter 4, [M+], enter 5, press [M+], enter 6 and press [M+]. Finally press [MR] to recall the memory. You will get 18.

Square roots

Use the key marked [√].

For example, to find the square root of 144, press 144 then press [√]. You will get 12.

Squares

This a number times itself. Use the key marked [x^2]. On some calculators this is a second function, which is shown on the calculator above another key.

For example, to find 67 squared, enter 67, then press [x^2]. You will get 4489.

Where the square is a second function, enter 67, then press the key marked [SHIFT] or [2nd F], then press the key above which x^2 is written.

Finding $\log_{10}$

This is log to the base ten and is a power function. It helps to simplify large numbers, for example for the purposes of scaling graphs. The key to use is usually marked [log].

For example, find the log of 10. Enter 10, press [log]. You will get 1.

To understand how the $\log_{10}$ function simplifies scaling, find logs of 10, 100, 1000, 10 000 and 100 000.

Note the relationship between the numbers.

$100 = 10^2$ and the log is 2; $1000 = 10^3$ and the log is 3.

Finding natural logs

This is useful for some biological calculations such as population estimation and species evenness.

The key to use is marked ln.

For example, to find the natural log of 46, enter 46, press ln and you will get 3.828.

Converting a number from $\log_{10}$ to base 10

If you have plotted a graph using logs, you may need to convert a value that you read off the graph back to base 10. The function you want to use is $10^{\square}$.

Press the key marked SHIFT then log. Enter the value you want to convert back to base 10 and press =.

For example, to find the value whose log is 3, press the key marked SHIFT then log. Enter the value 3 and press =. You will get 1000 (which is 10^3).

Converting a number from natural logs (ln) to base 10

If you have plotted a graph using natural logs (ln), you may need to convert a value that you read off the graph back to base 10. The function you want to use is $e^{\square}$.

Press the key marked SHIFT then ln. Enter the value you want to convert back to base 10 and press =.

For example, to find the value whose natural log is 6, press the key marked SHIFT then ln. Enter the value 6 and press =. You will get 403.4.

Using π

This is very easy because the value of π is a constant that is stored in the calculator. Simply press the key marked π and you will get 3.142. Note on many calculators this is a second function so you need to use the shift key.

Power functions

This means finding values for numbers such as 34^2 or 56^8. The key to use is marked y^x or x^y.

For example, to find the cube of 4 (4^3), first enter 4, then press y^x, then press the 3. You will get 64.

Reciprocals

The reciprocal of a number is 1 ÷ that number. The calculator key is marked $1/x$. This is a useful way of quickly finding values of $1/t$ in rates calculations.

For example, when $t = 56$ what is $1/t$? Enter 56, then press $1/x$.

You will get 0.018.

Standard form (powers of 10)

You can do calculations using standard form using the calculator. The key to use is marked EXP or EE or $\times 10^x$, depending on the model you have.

For example what is $(9 \times 10^6) \times (12 \times 10^5)$? Enter 9, press EXP, enter 6, press ×, enter 12, press EXP, enter 5, press =. You will get 1.08^{13} or 1.08×10^{13}.

Excel and calculations

Using Excel to help with calculations

It is possible to do calculations very quickly using Excel. This requires the use of simple formulae which, once mastered, can save you a lot of time as well as allow repeated calculation for several data sets. Once you are familiar with the basics you can move on and explore Excel to do more complex calculations.

WORKED EXAMPLE

On the Nelson Thornes website locate and open the file called 'Beetroot_cells_student_data', using Excel. (The table is shown at the bottom of page 112.)

You will find a table containing data for the percentage transmission of blue light by samples of leached beetroot pigment. There is data for seven temperatures and 16 repetitions.

How to calculate the means in the columns of data

1. Calculate the mean for the sixteen runs, using the spreadsheet tools. First, highlight the figures in the result columns, beginning with 63 top left in the 20 °C column and dragging to include all the figures, with 5 in the 80 °C column at bottom right. Then click the AutoSum tool $\sum$. You will get a total at the bottom of each column in a new cell. The figures should be 1191, 1095, 1106, 570, 249, 194 and 169.
2. Click the box under 16 in the run number column on the left (next to the cell with 1191 in it). Type 'total'.
3. In the next cell down type 'mean'
4. Select the box under 1191. Type '=' then click the cell with 1191, then type '/16'. This is a formula which causes the value in the 1161 box to be divided by 16. Check that the formula C20/16 appears in the tool bar. Press return and you should get the figure 74.4375
5. Highlight this cell. You will see a dark border with a black square marking the bottom right corner. Hover the cursor over this until you see a +, then press and hold the left mouse button and drag to highlight the boxes to the right under the remaining columns. Let go of the mouse button and the formula will automatically be pasted across and the remaining averages calculated. You will have a row of figures.
6. Highlight this row of averages and click the reduce decimal tool until you have one decimal place, giving a row of figures 74.4, 68.4, 69.1, 35.6, 15.5, 12.1, 10.6. These are the means for each column.

WORKED EXAMPLE

How to calculate the standard deviation for each mean

1. Select the cell under the one with mean written in it and type 'standard deviation'
2. Select the cell under the first average (i.e. under the number 74.4)
3. Type '=STDEV('
4. Click the top number in the set of data (63), hold down the left button and drag to highlight the whole column down to the bottom number (81)
5. Type')'
6. Press return
7. You will see a number, 10.53862
8. Check that the formula bar shows =STDEV(C4:C19)
9. Select the cell with the SD value in and use the bottom right black square to drag and highlight the cells to the right under the remaining averages. When you release the left button you will automatically paste and perform the calculation for the remaining data.

10 Highlight the SD values and use the reduce decimal button to get them to two decimal places. They should be 10.54, 11.23, 9.4, 10.10, 7.13, 5.26 and 4.10.

Have a quick look at the numbers in the data. The column for 30 °C has the largest value of SD and the numbers vary the most, while the column for 80 °C seems to be the most reliable.

WORKED EXAMPLE

How to use the standard deviation to calculate confidence intervals for graphs

Table of values of t when $p = 0.05$							
Degrees of freedom ($df = n - 1$)	$t_p(n-1)$	**df**	$t_p(n-1)$	**df**	$t_p(n-1)$	**df**	$t_p(n-1)$
1	12.71	11	2.2	21	2.08		
2	4.3	12	2.18	22	2.07		
3	3.18	13	2.16	23	2.07		
4	2.78	14	2.15	24	2.06	40	2.02
5	2.57	15	2.13	25	2.06	60	2
6	2.45	16	2.12	26	2.06	120	1.98
7	2.37	17	2.11	27	2.05	>120	1.96
8	2.31	18	2.1	28	2.05		
9	2.26	19	2.09	29	2.05		
10	2.23	20	2.09	30	2.04		

1 Work out the t value for your data using the table. For this example, with 16 values, df is 16 − 1 = 15. Check that you can find the t value for this number of degrees of freedom in the table and understand how. The t value you should have found is 2.13

2 Work out the square root of n. For your data, with 16 values in n, this will be 4. Check for yourself.

3 Go back to the Excel spreadsheet.

4 Select the cell under the one with standard deviation typed in it and type 'confidence interval'.

5 Select the cell under the S.D. value for the first column (below 10.54).

6 Type '=('

7 Click the box with the value of standard deviation for the column (here 10.54).

8 Type '/4)*2.13'

9 / means divide by; * means multiply.

10 The formula box in Excel should now read =(C22/4)*2.13

11 Press return and the value will be calculated; the number should be 5.611815

12 Select this cell, drag across to automatically calculate for the other columns.

13 Select all these values, use the reduce decimal to get 1 decimal place. The numbers should be 5.6, 6.0, 5.0, 5.4, 3.8, 2.8, 2.2

REMEMBER: Using Excel to do calculations is quite easy but it takes a bit of practise to become completely familiar.

SUMMARY QUESTION

1 Try using Excel spreadsheets to perform some of the calculations described earlier in this book.

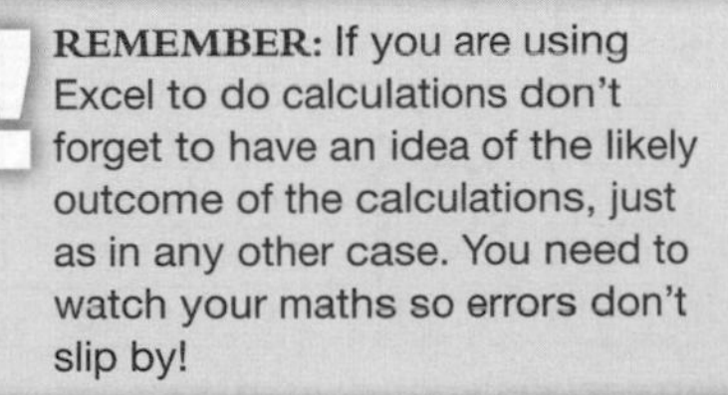

REMEMBER: If you are using Excel to do calculations don't forget to have an idea of the likely outcome of the calculations, just as in any other case. You need to watch your maths so errors don't slip by!

Table of standardised normal distribution

An entry in the table is the proportion under the entire curve which is between $z = 0$ and a positive value of z. Areas for negative values of z are obtained by symmetry.

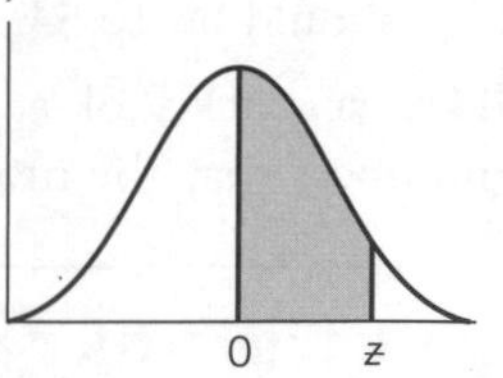

z	.00	.01	.02	.03	.04	.05	.06	.07	.08	.09
0.0	0.0000	0.0040	0.0080	0.0120	0.0160	0.0199	0.0239	0.0279	0.0319	0.0359
0.1	0.0398	0.0438	0.0478	0.0517	0.0557	0.0596	0.0636	0.0675	0.0714	0.0753
0.2	0.0793	0.0832	0.0871	0.0910	0.0948	0.0987	0.1026	0.1064	0.1103	0.1141
0.3	0.1179	0.1217	0.1255	0.1293	0.1331	0.1368	0.1406	0.1443	0.1480	0.1517
0.4	0.1554	0.1591	0.1628	0.1664	0.1700	0.1736	0.1772	0.1808	0.1844	0.1879
0.5	0.1915	0.1950	0.1985	0.2019	0.2054	0.2088	0.2123	0.2157	0.2190	0.2224
0.6	0.2257	0.2291	0.2324	0.2357	0.2389	0.2422	0.2454	0.2486	0.2517	0.2549
0.7	0.2580	0.2611	0.2642	0.2673	0.2703	0.2734	0.2764	0.2794	0.2823	0.2852
0.8	0.2881	0.2910	0.2939	0.2967	0.2995	0.3023	0.3051	0.3078	0.3106	0.3133
0.9	0.3159	0.3186	0.3212	0.3238	0.3264	0.3289	0.3315	0.3340	0.3365	0.3389
1.0	0.3413	0.3438	0.3461	0.3485	0.3508	0.3531	0.3554	0.3577	0.3599	0.3621
1.1	0.3643	0.3665	0.3686	0.3708	0.3729	0.3749	0.3770	0.3790	0.3810	0.3830
1.2	0.3849	0.3869	0.3888	0.3907	0.3925	0.3944	0.3962	0.3980	0.3997	0.4015
1.3	0.4032	0.4049	0.4066	0.4082	0.4099	0.4115	0.4131	0.4147	0.4162	0.4177
1.4	0.4192	0.4207	0.4222	0.4236	0.4251	0.4265	0.4279	0.4292	0.4306	0.4319
1.5	0.4332	0.4345	0.4357	0.4370	0.4382	0.4394	0.4406	0.4418	0.4429	0.4441
1.6	0.4452	0.4463	0.4474	0.4484	0.4495	0.4505	0.4515	0.4525	0.4535	0.4545
1.7	0.4554	0.4564	0.4573	0.4582	0.4591	0.4599	0.4608	0.4616	0.4625	0.4633
1.8	0.4641	0.4649	0.4656	0.4664	0.4671	0.4678	0.4686	0.4693	0.4699	0.4706
1.9	0.4713	0.4719	0.4726	0.4732	0.4738	0.4744	0.4750	0.4756	0.4761	0.4767
2.0	0.4772	0.4778	0.4783	0.4788	0.4793	0.4798	0.4803	0.4808	0.4812	0.4817
2.1	0.4821	0.4826	0.4830	0.4834	0.4838	0.4842	0.4846	0.4850	0.4854	0.4857
2.2	0.4861	0.4864	0.4868	0.4871	0.4875	0.4878	0.4881	0.4884	0.4887	0.4890
2.3	0.4893	0.4896	0.4898	0.4901	0.4904	0.4906	0.4909	0.4911	0.4913	0.4916
2.4	0.4918	0.4920	0.4922	0.4925	0.4927	0.4929	0.4931	0.4932	0.4934	0.4936
2.5	0.4938	0.4940	0.4941	0.4943	0.4945	0.4946	0.4948	0.4949	0.4951	0.4952
2.6	0.4953	0.4955	0.4956	0.4957	0.4959	0.4960	0.4961	0.4962	0.4963	0.4964
2.7	0.4965	0.4966	0.4967	0.4968	0.4969	0.4970	0.4971	0.4972	0.4973	0.4974
2.8	0.4974	0.4975	0.4976	0.4977	0.4977	0.4978	0.4979	0.4979	0.4980	0.4981
2.9	0.4981	0.4982	0.4982	0.4983	0.4984	0.4985	0.4985	0.4985	0.4968	0.4986
3.0	0.4987	0.4987	0.4987	0.4988	0.4988	0.4989	0.4989	0.4989	0.4990	0.4990

% transmission of blue light by samples of leached beetroot pigment							
Sheet 1							
Temperature/°C	**20**	**30**	**40**	**50**	**60**	**70**	**80**
run 1	63	47	63	19	7	19	15
2	82	79	74	52	31	11	10
3	85	86	80	23	11	8	12
4	82	77	71	38	13	13	18
5	78	73	69	25	21	20	7
6	82	81	85	35	19	8	17
7	75	63	63	39	19	22	13
8	69	72	64	50	22	14	9
9	78	72	61	45	22	12	10
10	52	57	58	41	10	8	4
11	65	63	64	39	9	17	7
12	71	60	61	35	20	14	7
13	86	65	87	24	20	4	10
14	57	51	67	31	10	10	11
15	85	81	80	47	10	8	14
16	81	68	59	27	5	6	5

Beetroot_cells_student_data